TED R. MUSGRAVE, Ph.D.
DEPARTMENT OF CHEMISTRY
COLORADO STATE UNIVERSITY

Understanding Problems For Chemical Principles

W. B. SAUNDERS COMPANY
Philadelphia London Toronto

1978

W. B. Saunders Company: West Washington Square
Philadelphia, PA 19105

1 St. Anne's Road
Eastbourne, East Sussex BN 21 3UN, England

1 Goldthorne Avenue
Toronto, Ontario M8Z 5T9, Canada

Library of Congress Cataloging in Publication Data

Musgrave, Ted R.

Understanding problems for chemical principles.

1. Chemistry, Physical and theoretical—Problems,
    exercises, etc.   I. Title.

QD456.M87 1978   540'.76   77–80749

ISBN 0–7216–6626–4

Cover photo courtesy of Jay M. Pasachoff, as illustrated in his book, *Contemporary
Astronomy*, published by W. B. Saunders Company, 1977.

Understanding Problems For Chemical Principles   ISBN 0–7216–6626–4

Last digit is the print number: 9   8   7   6   5   4   3   2   1

To Pat

PREFACE

"The stumbling way in which even the ablest of scientists in every generation have had to fight through thickets of erroneous observations, misleading generalizations, inadequate formulations and unconscious prejudice is rarely appreciated by those who obtain their scientific knowledge from textbooks."

. . . James Bryant Conant

At the beginning level of understanding the science of Chemistry, textbooks are the principal source of acquiring some knowledge of chemical principles. Regardless of the quality of a text, the competence of an instructor or the interest and enthusiasm of students, there always seems to be a need (by students) for "more help" with problems or "another source" to guide them toward an understanding of the complexities of chemistry problems. This little book was written to provide a student with some additional "tools" by which chemistry problems can become more understandable. The emphasis is on understanding, not just the *mechanics* of problem solving.

I have tried to put into writing some of the approaches to problems that I have found to be successful in the classroom at Colorado State University and the ·University of Hawaii. Throughout the book I have occasionally asked unanswered questions, made comments and generally tried to be informal.

Chapters are organized by categories. Sometimes this format of organization suggests artificial boundaries that do not, in fact, exist in chemistry. The format is used primarily to *aid* the student in sorting out the seemingly vast and often confusing *types* of problems encountered in a beginning chemistry course. The book contains 445 problems. The emphasis lies not on sheer numbers of problems but on explanations and *brief* discussion of principles underlying the problems. The sequence of material and general tone of the book are designed to coincide with the 4th Edition of *Chemical Principles* by Masterton and Slowinski.

I extend my appreciation to Dr. W. L. Masterton of the University of Connecticut for his thorough review of the manuscript and his many helpful suggestions. Even Dr. Masterton's most critical comments were a pleasure to read! My thanks to Jackie Swinehart for her invaluable help in preparing the manuscript and to Marilyn Bain-Ackerman for checking the problem solutions.

Ted R. Musgrave

CONTENTS

CONTENTS

1 THE FACTOR-LABEL METHOD

The factor-label method, variously called dimensionanalysis, unity-factor, or unit conversion, is a convenient problem solving technique that can be applied to a variety of chemistry problems and, indeed, to many "everyday" practical arithmetic problems. The method involves two steps: (1) stating the problem as a mathematical equation, and (2) multiplying the right side of the equation by conversion (unity) factors until the units on the right side of the equation match the units on the left.

With a current national interest in conversion to the metric system, a good way to introduce the factor-label method is to apply it to some typical metric conversions.

Example 1.

How many centimeters are there in one foot?

Solution.

First, we must state the problem as an equation. When you first attempt to use the factor-label method, this is often the most difficult step. We must set something "equal to" or "equivalent to" something else, and to do this usually requires some mental translation of the way the problem is stated. Another way of stating the problem here would be to ask, "How many centimeters are equivalent to one foot?" The unknown, "how many," we can abbreviate by calling it "X," and we have our equation

$$X \text{ cm} = 1.00 \text{ ft}$$

Next, we must multiply the right side of our equation by suitable conversion factors until "ft" becomes "cm." We must know some numerical conversion for length between the English and metric systems. If we knew that 1 foot = 30.48 cm, that would be fine, and in fact would be the answer to our problem, but let us suppose we did not remember that particular conversion, but we did remember that 1 inch = 2.54 cm. This will serve nicely as a conversion factor, or let's call it a "unity" factor. Unity simply means the number one; any conversion factor can be expressed as the number one. Thus, if we take the expression 1 inch = 2.54 cm and divide both sides by 1 inch, we have

$$\frac{1 \text{ inch}}{1 \text{ inch}} = \frac{2.54 \text{ cm}}{1 \text{ inch}}$$

or

$$1 = \frac{2.54 \text{ cm}}{1 \text{ inch}}$$

and if we divide both sides by 2.54 cm,

$$\frac{1 \text{ inch}}{2.54 \text{ cm}} = \frac{2.54 \text{ cm}}{2.54 \text{ cm}} = 1$$

Similarly, we know that 12 inches = 1 foot, so

$$\frac{12 \text{ inches}}{1 \text{ foot}} = 1, \text{ or } \frac{1 \text{ foot}}{12 \text{ inches}} = 1$$

In the factor-label method, it makes sense to think of conversion factors as unity factors. The reason is that we start with an equation, such as

$$X \text{ cm} = 1 \text{ ft}$$

and then proceed to multiply one side of the equation by conversion factors. Mathematically, in order for the equation to remain true, those conversions must be equal to one. We can multiply one side of an equation by 1 an infinite number of times without changing the original equation! For our problem, we can convert feet to inches by using the unity factor 12 inches/1 foot = 1. Thus,

$$X \text{ cm} = 1.00 \, \cancel{ft} \times \frac{12 \text{ in}}{1 \, \cancel{ft}}$$

The units of ft cancel out. We can then convert to cm using the unity factor 2.54 cm/1 in = 1.

$$X \text{ cm} = 1.00 \, \cancel{ft} \times \frac{12 \, \cancel{in}}{1 \, \cancel{ft}} \times \frac{2.54 \text{ cm}}{1 \, \cancel{in}}$$

The units on the right side now match the units on the left and the problem is solved, except for the arithmetic. Multiplying 12 × 2.54, the answer is 30.5 cm. Once the solution "set up" is correct, the multiplication and division may be accomplished quickly using a pocket calculator or slide rule. Notice that a little inspection (and practice) will show us which way to use a unity factor. Thus, 2.54 cm/1 in = 1, and 1 in/2.54 cm = 1, and in this particular problem we use it so that inches are in the denominator and will cancel to leave the desired units, cm, in the numerator.

Example 2.

How many grams are in one ton?

Solution.

Stated as an equation, we can "ask":

$$X \text{ g} = 1.0 \text{ ton}$$

We need an English to metric conversion for mass, such as 1 pound = 454 grams. In fact, all English to metric conversions can be carried out by the factor-label method if one remembers three conversion factors. One for length, e.g., 1 in = 2.54 cm; one for mass, 1 lb = 454 g; and one for volume, such as, 1 liter = 1.06 quarts. This problem can be solved by using two unity factors:

$$X \text{ g} = 1.0 \, \cancel{\text{ton}} \times \frac{2000 \, \cancel{\text{lbs}}}{1.0 \, \cancel{\text{ton}}} \times \frac{454 \text{ g}}{1 \, \cancel{\text{lb}}}$$

Units of tons cancel to give lbs, then units of lbs cancel to give the desired units of grams. The numerical answer is left to you.

Example 3.

How many milliliters of gasoline are there in a metric and in a U. S. gallon of gasoline?

Solution.

A metric gallon is 4.0 liters; hence there are 4000 ml in a metric gallon. For a U. S. gallon, we can use factor-label:

$$X \text{ ml} = 1 \text{ gal} \times \frac{4 \text{ qt}}{1.0 \text{ gal}} \times \frac{1.0 \text{ liter}}{1.06 \text{ qt}} \times \frac{1000 \text{ ml}}{1.0 \text{ liter}} = 3774 \text{ ml}$$

Example 4.

In 1976, the world's record in the 100 meter dash was 9.9 seconds. To accomplish this feat, what would be the athlete's average velocity in miles per hour?

Solution.

The question may be translated to "How many miles per hour are equivalent to 100 meters per 9.9 seconds?" Since "per" just means "divided by," the problem may be stated mathematically as

$$X \frac{\text{miles}}{\text{hour}} = \frac{100 \text{ meters}}{9.9 \text{ seconds}}$$

In this type of problem, there are units in both numerator and denominator that must be converted. It is most convenient to convert them one at a time. In our problem, let's operate on the numerator first and convert meters to miles, then convert seconds to hours in the denominator.

If we knew that 1 kilometer (1000 meters) = 0.62 miles, the former conversion could be accomplished in one step. But let's assume we know only that 1 in = 2.54 cm. Then,

$$X \frac{\text{mi}}{\text{h}} = \frac{100 \text{ m}}{9.9 \text{ s}} \times \frac{100 \text{ cm}}{1 \text{ m}} \times \frac{1 \text{ in}}{2.54 \text{ cm}} \times \frac{1 \text{ ft}}{12 \text{ in}} \times \frac{1 \text{ mi}}{5280 \text{ ft}}$$

Successively, units of m, cm, in, and ft cancel out until we are left with units of miles in the numerator. If we stopped at this point, the "answer" would be in miles per second, so we must next convert the denominator to hours:

$$X \frac{\text{mi}}{\text{h}} = \frac{100 \text{ m}}{9.9 \text{ s}} \times \frac{100 \text{ cm}}{1 \text{ m}} \times \frac{1 \text{ in}}{2.54 \text{ cm}} \times \frac{1 \text{ ft}}{12 \text{ in}} \times$$

$$\frac{1 \text{ mi}}{5280 \text{ ft}} \times \frac{60 \text{ s}}{1 \text{ min}} \times \frac{60 \text{ min}}{1 \text{ h}}$$

We now have the units we want, $\frac{\text{mi}}{\text{h}}$. The numerical answer is 22.6 mph. The world record in the 100 yard dash is 9.0 seconds. How many miles per hour is that? The record in the 1000 meters is 1 minute, 43.7 seconds. What would that average in mph? (ans. 21.5 mph)

Example 5.

A graphite pencil signature weighs 1.0 milligram. How many carbon atoms are there in the signature?

Solution.

The "question" equation is

$$X \text{ C atoms} = 1.0 \text{ mg C}$$

For the appropriate conversion factors, we need a basic concept of chemistry, one which you may not have covered at this point. That is, that there are 6.02×10^{23} atoms in 12.0 g of carbon: $12.0 \text{ g C} = 6.02 \times 10^{23}$ C atoms. The problem may now be solved as follows:

$$X \text{ C atoms} = 1.0 \text{ mg C} \times \frac{1 \text{ g C}}{1000 \text{ mg C}} \times \frac{6.02 \times 10^{23} \text{ C atoms}}{12.0 \text{ g C}} =$$

$$5.0 \times 10^{19} \text{ atoms}$$

Example 6.

The velocity of light is 3.00×10^{10} cm per second. How many miles does light travel in one year (i.e., one light-year)?

Solution.

Examine the solution step by step. It is lengthy but not difficult.

$$X \text{ mi} = 1 \text{ yr} \times \frac{365 \text{ d}}{1 \text{ y}} \times \frac{24 \text{ h}}{1 \text{ d}} \times \frac{60 \text{ min}}{1 \text{ h}} \times \frac{60 \text{ s}}{1 \text{ min}} \times$$

$$\frac{3.00 \times 10^{10} \text{ cm}}{1 \text{ s}} \times \frac{1 \text{ in}}{2.54 \text{ cm}} \times \frac{1 \text{ ft}}{12 \text{ in}} \times \frac{1 \text{ mi}}{5280 \text{ ft}} = 5.88 \times 10^{12} \text{ miles}$$

or 5,880,000,000,000 miles!

PROBLEMS

1. How many yards are there in 100 meters?

2. How many kilograms are in a 5 lb bag of sugar?

3. The density of mercury is 13.6 g per ml. What is the weight (in lbs) of one quart of mercury?

4. The distance to the sun is 93 million miles and light travels at 3.0×10^{10} cm per second. How long does it take for light from the sun to reach the earth?

5. How long would it take an automobile averaging 55 mph to travel 200 kilometers?

6. The artificial element californium 252 sells for $100 per tenth of a microgram (10^{-7} g). If one pound of ^{252}Cf were for sale, what would it cost?

7. A metric wrench has a span of 20 millimeters. How many inches is that?

8. If a carbon dioxide molecule were 5.0 angstroms in length, how many CO_2 molecules laid end to end would it take to reach 1 mile? (one angstrom = 10^{-8} cm)

9. Convert 55 miles per hour to meters per second.

10. What would be the weight (in lbs) of 100 billion atoms of lead? (at. wt. 207)

2 MOLES, FORMULAS, AND STOICHIOMETRY

The term stoichiometry refers to relationships between quantities of reactants and products in chemical systems. Formulas tell us the relative numbers of different atoms making up a chemical compound. One of the most important quantitative concepts in chemistry is the *mole*. Since individual atoms or molecules are so small as to be virtually impossible to work with, the idea of a large collection of atoms or molecules, namely the "mole," was developed. The mole is defined as the number of carbon atoms in 12.000 grams of the isotope $^{12}_6C$ – ordinary carbon. That number, known as *Avogadro's Number*, is 6.023×10^{23}. Strictly speaking, one mole of anything – atoms, molecules, ions, stars, grains of sand, etc. – is defined as 6.023×10^{23} of that thing. This is a huge number! One mole of marbles, for example, would completely cover the surface of 250 planets the size of the earth to a depth of five feet.

The *atomic weight* listed on periodic tables gives the mass of an atom of an element as it occurs in nature relative to that of a ^{12}C atom, which is taken to be exactly 12. Atomic weights are average values. Often they are not whole numbers because most natural elements consist of more than one isotope. Carbon, for example, is mostly $^{12}_6C$ but contains small amounts of $^{13}_6C$ and $^{14}_6C$, so the average atomic weight of carbon is 12.01115. The atomic weight of hydrogen is 1.00797; of iron, 55.847, and so on. While these numbers are often known to six or seven significant figures, the chemist usually rounds off to three or four figures. The *gram atomic weight* is the mass of an element, in grams, which is numerically equal to its atomic weight.

Molecular weight may be determined by summing atomic weights when the formula of a molecular substance is known. Thus, if we know that the formula of water is H_2O, this tells us that each water molecule contains two hydrogen atoms and one oxygen atom. One mole of water molecules would therefore contain two moles of hydrogen atoms and one mole of oxygen atoms. The molecular weight of H_2O is $(2 \times 1.01) + (1 \times 16.00) = 18.02$. The *gram molecular weight* of water is 18.02 grams.

Formula weight is often used instead of molecular weight. Strictly speaking, the term "molecular weight" should be used only with a molecular substance. Ionic compounds, such as NaCl, do not consist of molecules. Therefore, it is incorrect to refer to the "molecular weight" of sodium chloride, but it is perfectly accurate to say that the formula weight of NaCl is $23.0 + 35.5 = 58.5$. One mole of sodium chloride weighs 58.5 grams.

Problem Categories

Problems in this chapter are presented in three categories:

I   Conversions between grams, atoms, moles, and molecules.

II   Relationship of chemical formulas to elementary composition.

III Mass relationships in chemical equations.

CATEGORY I   **CONVERSIONS BETWEEN GRAMS, ATOMS, MOLES, AND MOLECULES**

Example 1.

How many moles of sulfur atoms are there in 8.23 grams of sulfur, S?

Solution.

One may convert from grams to moles by using the factor-label method. Here, we make use of the unity factor 1 mole S atoms = 32.06 grams S. Thus,

$$X \text{ moles S} = 8.23 \text{ grams S} \times \frac{1 \text{ mole S (atoms)}}{32.06 \text{ grams}} = 0.257 \text{ mole}$$

Example 2.

How many iron atoms are there in 1.00 gram of iron, Fe?

Solution.

Solving the problem requires two steps. First, convert grams to moles, then use Avogadro's number as a conversion factor since 1 mole of Fe atoms is equivalent to 6.02×10^{23} Fe atoms. Thus,

$$X \text{ Fe atoms} = 1.00 \text{ g Fe} \times \frac{1 \text{ mole Fe (atoms)}}{55.85 \text{ g Fe}} \times \frac{6.02 \times 10^{23} \text{ Fe atoms}}{1 \text{ mole Fe (atoms)}}$$

$$= 1.08 \times 10^{22} \text{ Fe atoms}$$

Example 3.

How many moles of carbon dioxide, CO_2, are there in 20.0 grams of CO_2?

Solution.

The molecular weight of CO_2 is 12.01 + 16.00 + 16.00 = 44.01 grams; thus,

$$1 \text{ mole } CO_2 = 44.01 \text{ g } CO_2$$

By the factor-label method:

$$X \text{ moles } CO_2 = 20.0 \text{ g } CO_2 \times \frac{1 \text{ mole } CO_2}{44.01 \text{ g } CO_2} = 0.454 \text{ mole}$$

Example 4.

How many sugar molecules are there in 5.00 grams of sugar, $C_{12}H_{22}O_{11}$?

Solution.

The solution is similar to Example 2, except that we now have molecules instead of atoms. The molecular weight of $C_{12}H_{22}O_{11}$ is $(12 \times 12.01) + (22 \times 1.01) + (11 \times 16.00) = 342.3$ grams.

$$X\ C_{12}H_{22}O_{11}\ \text{molecules} = 5.00\ \text{g}\ C_{12}H_{22}O_{11} \times \frac{1\ \text{mole}\ C_{12}H_{22}O_{11}}{342.3\ \text{g}\ C_{12}H_{22}O_{11}}$$

$$\times \frac{6.02 \times 10^{23}\ C_{12}H_{22}O_{11}\ \text{molecules}}{1\ \text{mole}\ C_{12}H_{22}O_{11}} = 8.79 \times 10^{21}\ \text{molecules}$$

Example 5.

If a box contained 9,000,000,000,000 oxygen molecules, O_2, how many moles of oxygen would that be?

Solution.

$$X\ \text{moles}\ O_2 = 9 \times 10^{12}\ \text{molecules}\ O_2 \times \frac{1\ \text{mole}\ O_2}{6.02 \times 10^{23}\ \text{molecules}\ O_2}$$

$$= 1.5 \times 10^{-11}\ \text{mole}$$

RELATIONSHIP OF CHEMICAL FORMULAS TO ELEMENTARY COMPOSITION

CATEGORY II

Example 1.

How many grams of carbon and of chlorine are there in 0.50 mole of carbon tetrachloride, CCl_4?

Solution.

From the formula CCl_4, we know that each mole of carbon tetrachloride contains one mole of carbon atoms and four moles of chlorine atoms. Furthermore, we know that 1 mole C (atoms) = 12.01 g C (atoms), so by factor-label

$$X\ \text{g C} = 0.50\ \text{mole}\ CCl_4 \times \frac{1\ \text{mole C (atoms)}}{1\ \text{mole}\ CCl_4\ \text{(molecules)}} \times \frac{12.01\ \text{g C (atoms)}}{1\ \text{mole C (atoms)}} = 6.01\ \text{grams}$$

Since there are four moles of chlorine atoms per mole of CCl_4, and the atomic weight of chlorine is 35.45, by factor-label

$$X\ \text{g Cl} = 0.50\ \text{mole}\ CCl_4 \times \frac{4\ \text{moles Cl}}{1\ \text{mole}\ CCl_4} \times \frac{35.45\ \text{g Cl}}{1\ \text{mole Cl}} = 70.90\ \text{grams}$$

Example 2.

How many grams of sulfur are there in 10.0 grams of sulfur trioxide, SO_3?

Solution.

From the formula we see that one mole of SO_3 (molecules) contains one mole of S (atoms). Solving by the factor-label method,

$$X\ \text{g S} = 10.0\ \text{g}\ SO_3 \times \frac{1\ \text{mole}\ SO_3}{80.1\ \text{g}\ SO_3} \times \frac{1\ \text{mole S}}{1\ \text{mole}\ SO_3} \times \frac{32.1\ \text{g S}}{1\ \text{mole S}} = 4.01\ \text{g}$$

The second unity factor may be eliminated by direct use of the relationship: 32.1 g S (atoms) is equivalent to 80.1 g SO_3 (molecules). Thus, we could have written

$$X \text{ g S} = 10.0 \text{ g } SO_3 \times \frac{32.1 \text{ g S}}{80.1 \text{ g } SO_3} = 4.01 \text{ g}$$

Example 3.

How many chlorine atoms are in 0.20 gram of DDT, $C_{14}H_9Cl_5$?

Solution.

There are 5 moles of Cl atoms per mole of DDT, and the molecular weight of DDT is $(14 \times 12.01) + (9 \times 1.01) + (5 \times 35.45) = 354.5$ g.

$$X \text{ Cl atoms} = 0.20 \text{ g DDT} \times \frac{1 \text{ mole DDT}}{354.5 \text{ g DDT}} \times \frac{5 \text{ moles Cl}}{1 \text{ mole DDT}} \times$$

$$\frac{6.02 \times 10^{23} \text{ Cl atoms}}{1 \text{ mole Cl}} = 1.70 \times 10^{21} \text{ atoms}$$

Example 4.

A sample of a chemical compound was analyzed and found to contain 5.60 grams of nitrogen and 12.80 grams of oxygen. What is the formula of the compound?

Solution.

Formulas are written to correspond to the relative numbers of atoms in molecules, ions, etc. Formulas of chemical compounds are determined from the ratios of the number of moles of the various atoms composing the compound. We know that this compound consists of N and O and will have a formula N_zO_y. The problem is to find z and y. From analysis, we know there are 5.60 g of N and 12.80 g of O. The first step is to convert to moles of N and O.

$$X \text{ moles N} = 5.60 \text{ g N} \times \frac{1 \text{ mole N}}{14.0 \text{ g N}} = 0.40 \text{ mole N}$$

$$X \text{ moles O} = 12.80 \text{ g O} \times \frac{1 \text{ mole O}}{16.0 \text{ g O}} = 0.80 \text{ mole O}$$

Since analysis of a compound is performed on some arbitrary amount, it would be coincidental if these numbers came out exactly 1, 2, 3, etc. They are usually always non-integers. Compounds cannot be made up of fractions of atoms, so we must convert the ratio of atoms to whole numbers. This is done by dividing the larger number(s) by the smallest:

$$\frac{0.80}{0.40} = \frac{2}{1} = \frac{\text{moles O atoms}}{\text{moles N atoms}}$$

The formula of the compound is NO_2.

The formula we have found, NO_2, is called the *simplest formula*. As the name implies, it tells us the simplest whole number ratio of atoms in the compound. However, this does not necessarily mean that the compound consists of molecules of NO_2. The compound may be ionic rather than molecular. Some information other than elemental analysis would be needed to determine the chemical nature of the compound. Even if the compound "NO_2" were known to be molecular, we would not know for sure if the molecules consisted simply of one nitrogen atom and two oxygen atoms, or two nitrogen atoms and four oxygen atoms, etc. All we know is that the ratio of N to O is 1:2. More information would be needed before we could ascertain the *molecular formula*.

Example 5.

Experiment revealed that the compound in Example 4 was molecular and had a molecular weight of 138.0. What is the molecular formula of the compound?

Solution.

If the simplest and molecular formulas were identical, i.e., NO_2, what would be the molecular weight? $(14.0) + (2 \times 16.0) = 46$. Obviously, the molecules must consist of more than one atom of N and two of O. What if the molecules were N_2O_4? The molecular weight would be: $(2 \times 14.0) + (4 \times 16.0) = 92$. How about N_3O_6? The molecular weight would be: $(3 \times 14.0) + (6 \times 16.0) = 138$. Bingo! The molecular formula is N_3O_6.

Thus, if the molecular weight of the compound is known, the molecular formula can be determined by dividing the true molecular weight by the simplest formula weight. In this case, $138 \div 46 = 3$ shows us that the molecular formula is three times the simplest formula, NO_2, or N_3O_6.

Example 6.

Analysis of a compound gave the following composition: 1.59% H, 22.22% N, and 76.19% O. Calculate the simplest formula of the compound.

Solution.

Chemical analyses are usually reported as per cent (by weight). How do you find moles from per cent? Choose any number of grams of compound you wish, then knowing per cent by weight, you can find the number of grams of a particular element in a known weight of that compound. Thus, in our example, one might arbitrarily choose 20.0 grams of the compound. How much of that is hydrogen? 1.59% of 20 g = 0.318 g. The easy way, however, is to choose 100.0 grams. Then 1.59%, 22.22%, and 76.19% would be 1.59, 22.22, and 76.19 grams of H, N, and O respectively. Now we can convert to moles.

$$X \text{ moles H} = 1.59 \text{ g H} \times \frac{1 \text{ mole H}}{1.01 \text{ g H}} = 1.57 \text{ moles H}$$

$$X \text{ mole N} = 22.22 \text{ g N} \times \frac{1 \text{ mole N}}{14.0 \text{ g N}} = 1.59 \text{ moles N}$$

$$X \text{ mole O} = 76.19 \text{ g O} \times \frac{1 \text{ mole O}}{16.0 \text{ g O}} = 4.76 \text{ moles O}$$

The atomic weights were rounded off to three significant figures, but the mole ratio of N/H is obviously 1/1. The ratio of O/H, or O/N, is: $4.76/1.59 = 3/1$. The simplest formula, then, is: $H_1N_1O_3$, or HNO_3.

MASS RELATIONSHIPS IN CHEMICAL REACTIONS   CATEGORY

III

In the examples in Categories I and II we have examined various conversions between grams, atoms, molecules, and moles and related these quantities to chemical formulas.

Now, let's turn our attention to what is called stoichiometry, that is, how these quantities relate to chemical equations. A chemical equation is an abbreviated means of quantitatively describing what happens in a chemical reaction. For example, molecules of methane and oxygen react to give molecules of carbon dioxide and water. This statement can be abbreviated by using molecular formulas and writing

$$CH_4 + O_2 \rightarrow CO_2 + H_2O$$

We can call this a "reaction expression," but it is not (yet) a chemical equation. In chemistry, as in mathematics, an equation must be "balanced." In the course of a chemical reaction, atoms cannot magically appear or disappear. We must end up with the same number and kind that we started with, regardless of how they become rearranged in the reaction. So, we must balance the numbers of atoms in a chemical reaction. Looking at the reaction above, the methane molecule contains 4 hydrogen atoms — what happens to them? They end up as part of water molecules. Yet, each water molecule contains only 2 hydrogen atoms, so to account for the hydrogens, it must be that two H_2O molecules are formed for each CH_4 that reacts. We should change the reaction to read

$$CH_4 + O_2 \rightarrow CO_2 + 2 H_2O$$

So far, so good! We've now indicated that one molecule of CH_4 gives two molecules of H_2O. That takes care of balancing the hydrogen, but what about the other atoms? Inspection shows one carbon atom on each side, so the carbon is balanced, but the oxygen is not. As the expression stands, there are two O atoms on the left and four on the right. The oxygens can be balanced by using two O_2 molecules on the left:

$$CH_4 + 2 O_2 \rightarrow CO_2 + 2 H_2O$$

We could also add subscripts to indicate whether the species were gas (g), liquid (l), or solid (s), and write, for example,

$$CH_4(g) + 2 O_2(g) \rightarrow CO_2(g) + 2 H_2O(l)$$

Now we have a balanced chemical equation that tells us that one CH_4 molecule combines with two O_2 molecules to produce one molecule of CO_2 and two molecules of H_2O. On a mole scale, we can say that one mole of CH_4 would react with two moles of O_2 to give one mole of CO_2 and two moles of H_2O.

Example 1.

Propane, C_3H_8, is commonly used as a fuel source. It burns (reacts with O_2) to produce CO_2 and H_2O. How many moles of O_2 are required to react with 0.50 mole of C_3H_8?

Solution.

The reaction $C_3H_8 + O_2 \rightarrow CO_2 + H_2O$ may be balanced by inspection to give the equation

$$C_3H_8 + 5 O_2 \rightarrow 3 CO_2 + 4 H_2O$$

In stoichiometry problems, the balanced equation is used to give the necessary unity factor(s). For this problem, we see that "5 moles of O_2 is equivalent to one mole of C_3H_8," and the problem may be solved:

$$X \text{ moles } O_2 = 0.50 \text{ mole } C_3H_8 \times \frac{5 \text{ moles } O_2}{1 \text{ mole } C_3H_8} = 2.5 \text{ moles } O_2$$

Example 2.

Green plants produce glucose, $C_6H_{12}O_6$, and molecular oxygen by "photosynthesis" of carbon dioxide and water. How many grams of CO_2 are required to produce one mole of glucose?

Solution.

The balanced equation is $6 \, CO_2 + 6 \, H_2O \rightarrow C_6H_{12}O_6 + 6 \, O_2$. Solving, we find

$$X \text{ g } CO_2 = 1 \text{ mole } C_6H_{12}O_6 \times \frac{6 \text{ moles } CO_2}{1 \text{ mole } C_6H_{12}O_6} \times \frac{44.0 \text{ g } CO_2}{1 \text{ mole } CO_2} = 264 \text{ g } CO_2$$

Example 3.

Using the equation in Example 1, calculate how many grams of water are produced when 10.0 grams of propane are burned.

Solution.

The equation tells us that 1 mole of C_3H_8 produces (is equivalent to) 4 moles of H_2O, and we must work through that relationship. Thus,

$$X \text{ g } H_2O = 10.0 \text{ g } C_3H_8 \times \frac{1 \text{ mole } C_3H_8}{44.0 \text{ g } C_3H_8} \times \frac{4 \text{ moles } H_2O}{1 \text{ mole } C_3H_8} \times \frac{18.0 \text{ g } H_2O}{1 \text{ mole } H_2O}$$
$$= 16.4 \text{ g } H_2O$$

It is necessary to convert from grams to moles (for C_3H_8); use the equation derived unity factor (4 moles H_2O = 1 mole C_3H_8), then convert back from moles H_2O to grams H_2O. A good rule for stoichiometry problems is to "think moles." Almost all stoichiometry problems must be solved by using a mole ratio obtained from a balanced equation.

Example 4.

How many moles of carbon dioxide, CO_2, could be produced from 2.0 moles of carbon and 3.0 moles of oxygen, O_2?

Solution.

The equation for the reaction is $C + O_2 \rightarrow CO_2$. We see that one mole of C would react with one mole of O_2 to give one mole of CO_2. What happens when we have 2 moles of C and 3 moles of O_2? The reaction will proceed until the carbon is consumed; then it will stop. One mole of O_2 will simply be left over. The two moles of C will combine with only two moles of O_2 and produce two moles of CO_2.

Example 5.

How many grams of water may be produced by igniting a mixture of 10.0 grams of hydrogen, H_2, and 10.0 grams of oxygen, O_2?

Solution.

We may write the balanced equation as $2 H_2 + O_2 \rightarrow 2 H_2O$. The point to note in this problem is that we do not start with an exact ratio of 2 moles of H_2 to 1 mole of O_2. There are 10.0 grams of H_2 and 10.0 grams of O_2 available. Let's convert to moles:

$$X \text{ moles } H_2 = 10.0 \text{ g } H_2 \times \frac{1 \text{ mole } H_2}{2.02 \text{ g } H_2} = 4.95 \text{ moles } H_2$$

$$X \text{ moles } O_2 = 10.0 \text{ g } O_2 \times \frac{1 \text{ mole } O_2}{32.0 \text{ g } O_2} = 0.313 \text{ mole } O_2$$

The mole ratio of $\dfrac{H_2}{O_2}$ is $\dfrac{4.95}{0.313} = \dfrac{15.8}{1}$. Obviously, there is a large excess of H_2 over and above that which is needed to just react with the O_2. The extent of the reaction will be limited by the amount of O_2 present, and when the oxygen is used up, the reaction will stop. So, we can re-phrase this problem to ask "How many grams of water may be produced from 0.313 mole of O_2?" Then, solving, we find

$$X \text{ g } H_2O = 0.313 \text{ mole } O_2 \times \frac{2 \text{ mole } H_2O}{1 \text{ mole } O_2} \times \frac{18.0 \text{ g } H_2O}{1 \text{ mole } H_2O} = 11.3 \text{ g } H_2O$$

PROBLEMS   *Category I*

1. How many moles are there in
   (a) 17.5 g of Sn?
   (b) 3.6 g of H_2O?
   (c) 50.0 g of NaCl?
   (d) 160.0 g of CCl_4?

2. Calculate the weight in grams of
   (a) 0.20 mole of methyl alcohol, CH_3OH.
   (b) 0.05 mole of phosphoric acid, H_3PO_4.
   (c) 1.0 mole of the amino acid histidine, $C_5H_6N_2O_2$.
   (d) 9.03×10^{23} molecules of ammonia, NH_3.

3. How many molecules are there in
   (a) 6.0 moles of benzene, C_6H_6?
   (b) 6.0 g of benzene?
   (c) 6.0 g of xenon tetrafluoride, XeF_4?
   (d) 6.0 g of NaCl? (Careful!)

4. Calculate the weight of
   (a) one lead atom.
   (b) 100 molecules of sugar, $C_{12}H_{22}O_{11}$.
   (c) 500 moles of Fe_2O_3.
   (d) 6.0 moles of NaCl.

Category II

5. How many moles of nitrogen atoms are there in 25.0 grams of dinitrogen pentoxide, N_2O_5?

6. How many grams of oxygen are in 4.0 grams of acetone, C_3H_6O?

7. How many moles of sulfur are there in 6.84 grams of $Al_2(SO_4)_3$?

8. If exactly 3.0 grams of a metal, M, combine with 6.0 grams of oxygen to give a compound having the formula MO_2, what is the atomic weight of metal M?

9. An oxide of nitrogen contains 30.4% nitrogen. What is the simplest formula of the compound, N_xO_y?

10. A compound consisting of Ca, Cr, and O was analyzed and found to contain 7.80 grams of Ca, 20.3 grams of Cr, and 21.84 grams of O. Calculate the simplest formula.

11. A compound that is 92.3% carbon and 7.7% hydrogen has a molecular weight of 78. What is the molecular formula of the compound?

12. Tetraethyl lead is $Pb(C_2H_5)_4$. What per cent lead (by weight) is this compound?

Category III

13. Balance the following reaction expressions:
    (a) $NO_2 + H_2O \rightarrow HNO_3 + NO$
    (b) $C_6H_6 + O_2 \rightarrow CO_2 + H_2O$
    (c) $Al_2O_3 + H_2O \rightarrow Al(OH)_3$
    (d) $C_3H_8O + O_2 \rightarrow CO_2 + H_2O$
    (e) $NH_3 + O_2 \rightarrow NO + H_2O$

14. Consider the reaction in 13 (a) above. How many moles of NO are produced from 13.8 grams of NO_2?

15. Ethyl alcohol reacts with phosphorus tribromide. The equation for the reaction is

$$3\,C_2H_5OH + PBr_3 \rightarrow 3\,C_2H_5Br + H_3PO_3$$

How many grams of C_2H_5Br could be produced from 5.42 grams of PBr_3?

16. A sample of a hydrocarbon compound, C_xH_y, burns in air to produce 8.80 grams of CO_2 and 5.40 grams of H_2O. What is the simplest formula of the compound?

17. Nitromethane, CH_3NO_2, is used as a fuel in aircraft and racing car engines. When burned the products are CO_2, H_2O, and NO. How many grams of oxygen are required to completely burn 122 grams of CH_3NO_2?

18. Molecular nitrogen, N_2, reacts with hydrogen, H_2, to produce ammonia, NH_3. How many grams of ammonia may be produced from a mixture of 50.0 grams of N_2 and 30.0 grams of H_2?

General Problems

19. Find the number of moles:
    (a) of He that weigh the same as 2.0 moles of Ne.
    (b) of gold in 1.0 lb of gold.
    (c) of stars in the Milky Way (150,000,000,000).
    (d) of silicon in 1.0 gram of sand (SiO_2).
    (e) of carbon in 10.0 grams of tetraethyl lead, $Pb(C_2H_5)_4$.

20. An unknown oxide of manganese is reacted with carbon to form manganese metal and carbon dioxide. Exactly 31.6 grams of the oxide, Mn_xO_y, yield 13.2 grams of CO_2. Find the simplest formula of the unknown oxide.

21. Nitrogen monoxide, NO, reacts with O_2 to form nitrogen dioxide, NO_2. How many grams of O_2 are consumed when 20.0 grams of NO_2 are formed by this reaction?

22. If a backyard charcoal broiler burns carbon at a rate of 2.0 grams per second, how many grams of carbon dioxide would it release into the atmosphere in 2 hours?

23. The heat transfer fluid used in most refrigerators is Freon 12, CCl_2F_2. It is produced by the reaction

$$CCl_4 + 2\ SbCl_4F \rightarrow CCl_2F_2 + 2\ SbCl_5$$

The antimony chlorofluoride used is made by the reaction

$$SbCl_5 + HF \rightarrow SbCl_4F + HCl$$

How many grams of HF are consumed in the production of 1000 grams of Freon 12?

24. Reaction of xenon with excess fluorine produced a mixture of three products, each containing Xe and F and weighing a total of 2.073 grams. By weight, the mixture contained 20.0% XeF_2, 60.0% A and 20.0% B. Product B contained 0.193 gram of fluorine. Identify A and B. (Total Xe = 1.331 g)

3 THERMODYNAMICS

The term "thermodynamics" literally means movement of heat. As it applies to chemistry, the term thermodynamics refers to all aspects of chemical systems that involve changes in energy of any type. With very few exceptions, chemical changes are always accompanied by energy changes.

The First Law of Thermodynamics defines the energy change for a process in terms of the difference between heat absorbed (Q) and work done (W). The Second Law of Thermodynamics puts a restriction on the amount of heat that may be converted into work and defines a quantity called entropy, S. A useful qualitative picture of entropy is that it is a measure of the tendency toward disorder in nature. The Second Law of Thermodynamics also leads to a quantity called the Gibbs Free Energy. The change in free energy, ΔG, for a process is given by: $\Delta G = \Delta H - T\Delta S$. The parameter, H, is called enthalpy and T is the absolute temperature. The enthalpy change, ΔH, is the heat flow for a process at constant pressure. The importance of ΔG is that it determines whether or not a change will occur spontaneously in a constant pressure system. If ΔG is negative, spontaneous change will occur, and if ΔG is positive, it will not. Any process in the physical universe, from the growth of a green plant to the explosion of a supernova, will be accompanied by a change in free energy. Naturally, the chemist is concerned with the ΔG values for chemical reactions.

A chemist may wish to know, "If I mix hydrogen gas and iodine, will they react to produce hydrogen iodide?" If the value of ΔG is known, or can be determined, the chemist can tell not only whether or not the reaction will occur (spontaneously) at a particular temperature and pressure, but also to what extent it will occur! The free energy change is dictated by two terms, ΔH, the enthalpy change, and $T\Delta S$, the entropy change term, or what is sometimes called the "unavailable energy" term. In this chapter, we will examine each of these terms as they apply to chemical change, beginning with the ΔH term. We will divide ΔH problems into three categories. A fourth category, which may be discussed later in your course, considers the role of entropy and free energy in thermodynamics.

Problem Categories

    I   Heats of reaction from bond energies.
   II Heats of reaction from heats of formation.
 III Experimental determination of heat energy change.
 IV The role of entropy and free energy.

HEATS OF REACTION FROM BOND ENERGIES

Energy may be considered an integral part of chemical reactions. In nearly all chemical reactions, energy is either absorbed or given off, and the energy is usually observed as heat energy. Where does the energy come from? Ignite a mixture of hydrogen and oxygen and what happens? Uncontrolled, the mixture explodes violently; controlled, it burns with a liberation of large quantities of heat. That energy doesn't magically appear out of the air. It is the result of the making and breaking of chemical bonds. If we inspect the simplest of molecules, H_2, we find it to be quite stable. So stable, in fact, that to break apart one mole (2.0 grams) of hydrogen molecules into hydrogen atoms required 103,000 calories of heat energy. One *calorie* is the heat energy needed to raise the temperature of one gram of water 1°C, so 103 kilocalories is a considerable amount of energy. Conversely, if we combined 2 moles of H atoms to give one mole of H_2, 103 kcal of energy would be released to the surroundings. That amount of energy, 103 kcal, is called the *bond energy* of H_2. Table 3–1 lists some representative bond energies, in kcal per mole, based on the gas state.

Table 3–1   Representative Bond Energies*

Bond		Energy (kcal)
H:H	H—H	103
Cl:Cl	Cl—Cl	57
O::O	O=O	116
N:::N	N≡N	225
O:H	O—H	109
N:H	N—H	92
C:H	C—H	98
H:Cl	H—Cl	102
Br:Br	Br—Br	45
I:I	I—I	36
H:I	H—I	71
C:Cl	C—Cl	78
H:Br	H—Br	87
C:C (Diamond)	C—C	85

*Most values listed are from R. Keller, Basic Tables in Chemistry, McGraw-Hill Company, 1967, p. 206.

While the table is by no means complete, it will suffice to illustrate the use of bond energies to calculate the heat energy changes in chemical reactions, and vice-versa.

Example 1.

From the table of bond energies, calculate the heat energy change for the gas phase reaction

$$H_2 (g) + \frac{1}{2} O_2 (g) \rightarrow H_2O(g)$$

Solution.

Let's consider the heat energy change per mole of H_2O produced. In order to produce one mole of H—O—H, we must break apart one mole of H—H and ½ mole of O=O molecules. Formation of O—H bonds releases energy, but the splitting of H_2 and O_2 bonds requires energy.

In this age, an analogy between energy and money seems quite appropriate, so let's consider the energy of a chemical reaction in terms of "cost" and "gain." It costs energy for the reactions

$$H_2 \rightarrow 2 H$$

$$\tfrac{1}{2} O_2 \rightarrow O$$

Energy is gained, i.e., *released*, from the reaction

$$2 H + O \rightarrow H—O—H$$

Our "bookkeeping" is as follows:

Cost

1 mole H—H bonds =	103 kcal	
$\tfrac{1}{2}$ mole O=O bonds =	58 kcal	($\tfrac{1}{2}$ × 116)
Total =	161 kcal	

Gain   2 moles of O—H bonds = 218 kcal (2 × 109)

The formation of H_2O from H_2 and O_2 looks like "good business in the energy game!" We get back more than we had to put in. The difference is 57 kcal, and that's pure profit!

Stated in somewhat more scientific terminology, there is a net release of 57 kcal of energy in the formation of one mole of H_2O (g) from H_2 (g) and O_2 (g). This is called the *"heat of the reaction,"* and, in this case, it is also called the *heat of formation* of H_2O (g), since it is the heat energy change when one mole of H_2O(g) is formed from the elements (H_2 and O_2). If we incorporate that 57 kcal in a chemical equation, the equation becomes

$$H_2 \text{ (g)} + \tfrac{1}{2} O_2 \text{ (g)} \rightleftarrows H_2O \text{ (g)} + 57 \text{ kcal}$$

By convention, when describing the ΔH for a reaction, a minus sign would be used to indicate that heat energy was released, and we would write:

$$\Delta H_{rex'n} = -57 \text{ kcal}$$

Example 2.

From the table of bond energies calculate the ΔH of reaction for

$$H_2 \text{ (g)} + Br_2 \text{ (g)} \rightarrow 2 \text{ HBr(g)}$$

Solution.

Using the bookeeping approach we used in Example 1, we find

Cost

1 mole H_2 =	103 kcal
1 mole Br_2 =	45 kcal
Total =	148 kcal

Gain   2 moles H—Br bonds = 2 × 87 = 174 kcal

The difference is 26 kcal. This is a net "gain," or energy *released*, and ΔH for the reaction is -26 kcal.

Example 3.

From bond energies, calculate ΔH for the reaction

$$CH_4 + 2\ Cl_2\ (g) \rightarrow CCl_4 + 2\ H_2\ (g)$$

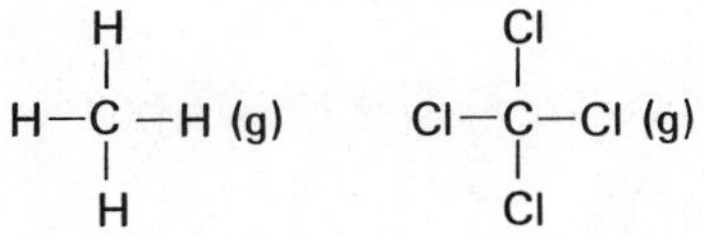

$$
\begin{array}{cccc}
 & H & & Cl \\
 & | & & | \\
H- & C & -H\ (g) & \quad Cl-C-Cl\ (g) \\
 & | & & | \\
 & H & & Cl
\end{array}
$$

Solution.

In this reaction, four moles of C—H bonds and two moles of Cl—Cl bonds must be broken, so the overall cost is

$$4 \times 98 = 392 \text{ kcal}$$

plus $\quad 2 \times 57 = \underline{114 \text{ kcal}}$

$$\text{Total} = 506 \text{ kcal}$$

How much energy do we get back? There are four moles of C—Cl bonds and two moles of H—H bonds formed, so the gain is

$$4 \times 78 = 312 \text{ kcal}$$
$$2 \times 103 = \underline{206 \text{ kcal}}$$
$$\text{Total} = 518 \text{ kcal}$$

Once again, we have an example of a reaction in which we get back more energy than we had to put in, and ΔH for the reaction is -12 kcal.

Example 4.

For the reaction $CH_4\ (g) + 4\ HI\ (g) \rightarrow CI_4\ (g) + 4\ H_2\ (g)$, ΔH is found to be +36 kcal. Estimate the bond energy of the C—I bond.

Solution.

From the table, we can find the total energy cost:

$$4 \times C-H = 4 \times 98 = 392 \text{ kcal}$$
$$4 \times H-I = 4 \times 71 = \underline{284 \text{ kcal}}$$
$$\text{Total} = 676 \text{ kcal}$$

Now, since the overall ΔH for the reaction is +36 kcal, we know that the energy cost *exceeds* the energy gain by 36 kcal. The total energy gain must be 640 kcal. Part of the energy gain is due to formation of four moles of H—H bonds, and part is due to forming the four moles of C—I bonds in CI_4. The contribution from the hydrogen is

$$4 \times H-H = 4 \times 103 = 412 \text{ kcal}$$

The balance (640 − 412 = 228 kcal) must be due to the CI_4 bonds. Thus, the C—I bond energy, per mole, is ¼ × 228 = 57 kcal.

The relation between heat of reaction and bond energies may be stated mathematically:

$$\Delta H \text{ (reaction)} = \text{(total bond energy cost)} - \text{(total bond energy gain)}$$

Thus, in this example, see that

$$+36 \text{ kcal} = (392 + 284) \text{ kcal} - (412 + 228) \text{ kcal}$$

HEATS OF REACTION FROM HEATS OF FORMATION   CATEGORY II

The heat of formation, ΔH_f, of a compound is defined as the heat of reaction when one mole of a compound is formed from its elements in their stable form. The heat of formation of an element, in its stable form, at 25°C and 1.0 atm pressure is zero. Heats of formation are experimental values and may be found to vary slightly depending upon the reference source. Table 3–2 lists some typical heats of formation. ΔH values are given in kcal per mole, at 25°C and 1 atm pressure.

Table 3–2   Representative Heats of Formation

Compound	ΔH	Compound	ΔH	Compound	ΔH
$H_2O(l)$	−68.4	$LiF(s)$	−146.3	$H_2SO_4(aq)$	−216.9
$H_2O(g)$	−57.8	$NaF(s)$	−136.0	$HNO_3(l)$	− 41.4
$H_2O_2(l)$	−44.8	$NaCl(s)$	− 98.2	$HNO_3(aq)$	− 49.4
$HF(g)$	−64.2	$NaBr(s)$	− 86.0	$H_3PO_4(aq)$	−308.2
$HCl(g)$	−22.1	$NaI(s)$	− 68.8	$H_3PO_3(aq)$	−232.2
$HBr(g)$	− 8.7	$CaO(s)$	−151.9	$CaCO_3(s)$	−288.5
$HI(g)$	+ 6.2	$BaO(s)$	−133.4	$BaCO_3(s)$	−291.3
$H_2S(g)$	− 4.8	$Al_2O_3(s)$	−399.1	$Na_2SO_4(s)$	−330.9
$NH_3(g)$	−11.0	$Cr_2O_3(s)$	−269.7	$NaHSO_4(s)$	−269.2
$CH_4(g)$	−17.9	$CO(g)$	− 26.4	$NaOH(s)$	−102.0
$C_2H_6(g)$	−20.2	$CO_2(g)$	− 94.1	$Ca(OH)_2(s)$	−235.8
$C_2H_4(g)$	+12.5	$SiO_2(s)$	−205.4	$CaSO_4(s)$	−342.4
$C_2H_2(g)$	+54.2	$SO_3(g)$	− 94.5	$CaSO_4\cdot2H_2O(s)$	−483.1

Notice that the heat of formation of a compound can differ with different physical states of the compound. Thus, the formation of gaseous H_2O from H_2 and O_2 is accompanied by the release of 57.8 kcal (per mole) of energy, but if the gaseous water condenses to the liquid state, an additional 10.6 kcal is given off, and the heat of formation of liquid water is listed as −68.4 kcal.

There is a direct relationship between the heat of a reaction and the heats of formation of the compounds involved in the reaction. This relationship is

$$\Delta H \text{ (reaction)} = \Sigma \, \Delta H_f \text{ (products)} - \Sigma \, \Delta H_f \text{ (reactants)}$$

Example 1.

Use the table of heats of formation to calculate the heat of reaction for

$$CaCO_3 (s) \rightarrow CaO (s) + CO_2 (g)$$

Solution.

If products are one mole of $CaO(s)$ and one mole of $CO_2(g)$ and the reactant is one mole of $CaCO_3(s)$, we find that

$$\Delta H \text{ (reaction)} = (\Delta H_f \text{ CaO} + \Delta H_f \text{ CO}_2) - (\Delta H_f \text{ CaCO}_3)$$

$$= (-151.9 + -94.1) - (-288.5)$$

$$= -246.0 + 288.5 = +42.5 \text{ kcal}$$

The decomposition of one mole of $CaCO_3(s)$ into $CaO(s)$ and $CO_2(g)$ requires 42.5 kcal of heat energy and is said to be *endothermic* (heat in). A reaction in which heat energy is given off (ΔH negative) is called *exothermic* (heat out).

Example 2.

Calculate ΔH (reaction) for

$$H_2S(g) + 2\,O_2(g) \rightarrow SO_3(g) + H_2O(g)$$

Solution.

Notice two things. One, the H_2O appearing in the reaction is H_2O gas and not H_2O liquid, so be careful to use the ΔH_f value for $H_2O(g)$. Secondly, one of the reactants is $O_2(g)$, and ΔH_f for oxygen in its natural state (O_2 gaseous) is zero. Solving,

$$\Delta H \text{ (reaction)} = (\Delta H_f \text{ SO}_3 + \Delta H_f \text{ H}_2\text{O(g)}) - (\Delta H_f \text{ H}_2\text{S})$$

$$= (-94.5 + -57.8) - (-4.8) = -147.5 \text{ kcal}$$

We find that the oxidation of H_2S is a highly exothermic reaction.

Example 3.

Calculate ΔH (reaction) for

$$Ca(OH)_2(s) + H_2SO_4(aq) \rightarrow CaSO_4(s) + 2\,H_2O(l)$$

starting with 37.04 g (0.5 mole) of $Ca(OH)_2(s)$ and 49.03 g (0.5 mole) of $H_2SO_4(aq)$.

Solution.

Note that H_2O appears as liquid water. Also, since ΔH_f values are listed in kcal/mole, we must divide the numbers found in the Table 3–2 by two when using only 0.5 mole of reactants. Solving, we find

$$\Delta H \text{ (reaction)} = (\Delta H_f \text{ CaSO}_4 + 2 \times \Delta H_f \text{ H}_2\text{O(l)}) -$$

$$(\Delta H_f \text{ Ca(OH)}_2 + \Delta H_f \text{ H}_2\text{SO}_4)$$

$$= (-171.2 + 2 \times -34.2) - (-117.9 + -108.5)$$

$$= (-239.6 + 226.4)$$

$$= -13.2 \text{ kcal}$$

Example 4.

Given the reaction:

$$Al_2O_3 \text{ (s)} + 6 \text{ HCl(g)} \rightarrow 2 \text{ AlCl}_3 \text{ (s)} + 3 \text{ H}_2O(g)$$

The reaction is endothermic with a heat of reaction of +83.7 kcal per mole of Al_2O_3 reacted. Find ΔH_f in kcal per mole for $AlCl_3$ (s).

Solution.

From Table 3–2 we find that the molar heats of formation of Al_2O_3 (s), HCl(g), and H_2O (g) are –399.1, –22.1, and –57.8 kcal, respectively. We can substitute into the equation and solve for ΔH_f $AlCl_3$:

$$\Delta H \text{ (reaction)} = (2 \times \Delta H_f + 3 \times \Delta H_f \text{ H}_2O) - (\Delta H_f \text{ Al}_2O_3 + 6 \times \Delta H_f \text{ HCl})$$

$$+83.7 = (2 \times \Delta H_f + -173.4) - (-399.1 + -132.6)$$

$$2 \times \Delta H_f \text{ AlCl}_3 = -274.6$$

$$\Delta H_f \text{ AlCl}_3 = -137.3 \text{ kcal}$$

EXPERIMENTAL DETERMINATION OF HEAT ENERGY CHANGE   CATEGORY III

Experimental determination of heats of reaction is called *calorimetry*, which means "heat measurement." A simple calorimeter is shown in Figure 3–1.

When an exothermic reaction occurs in the reaction chamber, the heat energy given off is released to the surroundings — the chamber walls and the water. The temperature of the water will increase, and from the rise in temperature, the number of calories of heat energy released may be determined. If a significant amount of heat is absorbed by the walls of the container, reaction chamber, etc., the calorimeter can be calibrated, using a known reaction, and its heat capacity determined. The *heat capacity* is the number of calories needed to raise the temperature by 1.0°C. In simple calorimetry experiments, the heat energy absorbed by material other than the water may be considered negligible, and the number of calories released by the reaction is determined by the heat capacity of the water alone.

Example 1.

A reaction is carried out in a calorimeter containing 100.0 g of water. As the reaction

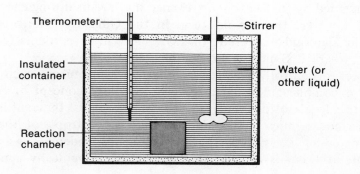

proceeds, the temperature of the water increases from 25.2°C to 27.2°C. Calculate the heat energy released by the reaction.

Solution.

Since one calorie is defined as the heat required to raise the temperature of one gram of water by 1.0°C, it follows that the heat capacity of 100.0 g of water is exactly 100.0 calories. Simple logic dictates that for the temperature of 100.0 g of water to increase by 2.0°C requires 200.0 calories, and that is the answer to the problem. Mathematically, we can say that

$$\Delta H = C \times \Delta t$$

Where ΔH is the heat change, C is the heat capacity of the calorimeter, and Δt is the change in temperature during reaction. Substituting the data from our problem, we see that

$$\Delta H = 100 \text{ cal/}°C \times 2.0°C$$

$$= 200 \text{ calories}$$

Example 2.

One gram of organic material from a sample of Colorado oil shale was burned in a "bomb" calorimeter. The temperature of the calorimeter increased from 25.0°C to 29.7°C. In a separate experiment it was found that it required 2690 calories to raise the temperature of the calorimeter by 1.0°C. Calculate the heat of combustion of the organic material.

Solution.

The heat capacity, C, of the calorimeter is 2690 cal./°C, and substituting into the equation used in Example 1, we find

$$\Delta H = C \times \Delta t$$

$$= 2690 \text{ cal/}°C \times 4.70°C$$

$$= 12,640 \text{ calories, or}$$

$$= 12.64 \text{ kcal}$$

CATEGORY IV   THE ROLE OF ENTROPY AND FREE ENERGY

The equation defining free energy, $\Delta G = \Delta H - T\Delta S$, is known as the Gibbs-Helmholtz equation. In chemical systems, at moderate temperatures and pressures, ΔG, and hence the driving force and spontaneity of chemical reaction, is usually governed by the heat energy change that occurs during chemical change. Sometimes, however, the second term in the Gibbs-Helmholtz equation will determine the sign and magnitude of ΔG. This will be particularly true at high temperature, where $T\Delta S$ may be large compared to ΔH.

The effect of entropy changes in physical phenomena, such as chemical reaction, is a complicated subject. Indeed, the very concept of entropy, as a measure of the degree of disorder, is somewhat nebulous! Inherent in the Second Law of Thermodynamics is the concept that the entropy of the universe is constantly increasing, i.e., that there is a spontaneous tendency toward *disorder* in nature. That, at least, is an idea that we can philosophically appreciate from personal experience. But, how does one "measure" entropy? We can measure heat energy in a calorimeter, but, as yet, there is no such thing as a "disorder meter." So, while entropy is important in chemical systems, it is a quantity that must be determined by indirect methods.

Values of free energy changes for chemical reactions can be determined by direct experimental methods (from equilibrium constants, standard potentials, etc.). The sign and magnitude of ΔG for chemical change is dependent upon temperature and pressure. In order to standardize quantitative information about ΔG values for chemical reactions, chemists have compiled tables of "free energies of formation," analogous to heats of formation discussed in Category II. The free energy of formation of a compound, ΔG_f, is the free energy change that occurs when one mole of a compound is produced from its elements (in their stable form) at 25°C and 1 atm pressure. Table 3–3 offers a partial listing of ΔG_f values, in units of kcal/mole.

Table 3–3   Representative Free Energies of Formation

Compound	ΔG_f	Compound	ΔG_f	Compound	ΔG_f
$H_2O(l)$	−56.7	$NaF(s)$	−129.3	$HNO_3(l)$	− 19.1
$H_2O(g)$	−54.6	$NaCl(s)$	− 91.8	$CaCO_3(s)$	−269.8
$HF(g)$	−64.7	$CaO(s)$	−144.4	$BaCO_3(s)$	−272.2
$HCl(g)$	−22.8	$BaO(s)$	−126.3	$CaSO_4(s)$	−315.6
$HBr(g)$	−12.7	$Al_2O_3(s)$	−376.8	$BaSO_4(s)$	−323.4
$H_2S(g)$	− 7.9	$Cr_2O_3(s)$	−250.2	$NO(g)$	+ 20.7
$NH_3(g)$	− 4.0	$CO(g)$	− 32.8	$NO_2(g)$	+ 12.4
$CH_4(g)$	−12.1	$CO_2(g)$	− 94.3	$PbO(s)$	− 45.1
$C_2H_6(g)$	− 7.9	$Ca(OH)_2(s)$	−214.3	$PbO_2(s)$	− 52.3
$C_2H_4(g)$	+16.3	$SO_3(g)$	− 88.5	$CCl_4(l)$	− 16.4

Like heat of formation, the free energy of formation of an element in its stable form is zero. Also, analogous to our discussion in Category II, the free energy change for a chemical reaction at 1 atm pressure and 25°C can be found from

$$\Delta G \text{ (1 atm)} = \Sigma\Delta G_f \text{ (products)} - \Sigma\Delta G_f \text{ (reactants)}$$

Example 1.

Find ΔG at 1 atm and 25° C for the reactions:

(a) $C_2H_4(g) + H_2(g) \rightarrow C_2H_6(g)$

(b) $CH_4(g) + 2 Cl_2(g) \rightarrow CCl_4(l) + 2 H_2(g)$

(c) $Cr_2O_3(s) + 3 CO(g) \rightarrow 2 Cr°(g) + 3 CO_2(g)$

Solution.

(a) ΔG (1 atm) = $(\Delta G_f\ C_2H_6(g)) - (\Delta G_f\ C_2H_4(g) + \Delta G_f\ H_2(g))$. Since $\Delta G_f\ H_2(g)$ is zero, the equation becomes

$$\Delta G \text{ (1 atm)} = (-7.9 \text{ kcal}) - (+16.3 \text{ kcal}) = -24.2 \text{ kcal}$$

We can conclude that, under the conditions specified, this reaction would occur spontaneously since ΔG is negative.

(b) ΔG (1 atm) = (ΔG_f CCl_4 (l)) − (ΔG_f CH_4 (g)). Note that ΔG_f for H_2 (g) and Cl_2 (g) are both zero.

$$\Delta G \text{ (1 atm)} = (-16.4 \text{ kcal}) - (-12.1 \text{ kcal}) = -4.3 \text{ kcal}$$

This reaction, too, should proceed spontaneously at 1 atm and 25°C, although the driving force for the reaction is small (only 4.3 kcal).

(c) ΔG (1 atm) = (3 × ΔG_f CO_2 (g)) − (ΔG_f Cr_2O_3 (s) + 3 × ΔG_f CO (g)). For $Cr°$ (s), chromium metal, ΔG_f is zero.

$$\Delta G \text{ (1 atm)} = (3 \times -94.3 \text{ kcal}) - (-250.2 \text{ kcal} + 3 \times -32.8 \text{ kcal})$$

$$= -282.9 \text{ kcal} + 348.6 \text{ kcal}$$

$$= +65.7 \text{ kcal}$$

We would conclude that carbon monoxide could not be used to reduce Cr_2O_3 to chromium metal at 25°C and 1 atm pressure.

Example 2.

It has been determined that the free energy of formation of H_2O (gas) is −54.6 kcal per mole and the free energy of formation of liquid water is −56.7 kcal per mole at 25°C and 1 atm pressure. Calculate the entropy change, ΔS, for the vaporization of water at 25°C and 1 atm.

Solution.

Since we know the free energies of formation, and heats of formation (see Table 3–2) for H_2O (l) and H_2O (g), we can calculate ΔG and ΔH for the change

$$H_2O(l) \rightarrow H_2O(g)$$

The heats of formation of H_2O (l) and H_2O (g) are −68.4 and −57.8 kcal, respectively. The change from liquid to gaseous water is endothermic with a ΔH value of

$$\Delta H = (-57.8) - (-68.4) = +10.6 \text{ kcal}$$

The free energy change is

$$\Delta G = (-54.6) - (-56.7) = +2.1 \text{ kcal}$$

Substituting into the Gibbs-Helmholtz equation at 25°C (298°K), we find

$$\Delta G = \Delta H - T\Delta S$$

$$+ 2.1 \text{ kcal} = +10.6 \text{ kcal} - (298°K) \times \Delta S$$

$$\Delta S = \frac{-8.5 \text{ kcal}}{-298°K} = +0.0285 \text{ kcal/°K}$$

Several ramifications of our solution to this problem should be noted. First, ΔH is positive. The reaction is endothermic and considerable heat energy is required to vaporize water. This quantity is called the "heat of vaporization" of water. Second, ΔG is positive. At first this seems surprising, since it means that water does not spontaneously vaporize at 25°C and 1 atm. We know that water doesn't spontaneously boil at 25°C and 1 atm pressure, but it does evaporate. The familiar process of evaporation at room temperature can be explained by the fact that under usual conditions, such as water in a dishpan, the water molecules can absorb heat energy from the surroundings and some can attain sufficient kinetic energy to escape from the liquid to the vapor phase. In a completely insulated, isoalted system, water does not spontaneously vaporize at 25°C and 1 atm. Third, notice that ΔS is positive. This results from the fact that water molecules are more disordered in the vapor phase than in the liquid. Any process that leads to a *greater degree of disorder* will have a *positive* ΔS.

So, while the positive ΔS term favors the spontaneous vaporization of water, the larger positive ΔH term dictates the sign on ΔG at 25°C. ΔH is +10.6 kcal and $T\Delta S$ is only +8.5 kcal, so ΔG is +2.1 kcal.

Example 3.

Assuming constant values of ΔH and ΔS, find the temperature at which water spontaneously vaporizes at 1 atm pressure.

Solution.

From Example 1, we found that both ΔH and ΔS were positive. Since ΔS is positive, there will be some temperature at which the $T\Delta S$ term becomes equal to the ΔH term. At this temperature, ΔG will equal zero, and at even higher temperatures, ΔG will be negative. All we need to do is substitute into the Gibbs-Helmholtz equation and solve for T at $\Delta G = 0$.

$$\Delta G = \Delta H - T\Delta S$$

$$0 = +10.6 \text{ kcal} - T \ (+0.0285 \text{ kcal/}^\circ K)$$

$$T = \frac{+10.6 \text{ kcal}}{+0.0285 \text{ kcal/}^\circ K} = 372^\circ K = 99^\circ C$$

The boiling point of water at 1 atm pressure is 100°C. Above this temperature, the transition from liquid to vapor is spontaneous, i.e., ΔG for the reaction is negative.

Example 4.

For the vaporization of solid NaCl at 25°C and 1 atm, ΔH and ΔS are +184 kcal and +0.0543 kcal/°K, respectively. For the hydration of gaseous Na^+ and Cl^- ions, ΔH and ΔS are -183 kcal and -0.0440 kcal/°K. Calculate ΔG for the solution of NaCl (NaCl(s) $\rightarrow Na^+(aq) + Cl^-(aq)$).

Solution.

Consider the process by steps:

$$NaCl(s) \xrightarrow{(1)} Na^+(g) + Cl^-(g) \xrightarrow{(2)} Na^+(aq) + Cl^-(aq)$$

Step (1) is endothermic with ΔH = +184 kcal and step (2) is exothermic with a ΔH of -183 kcal. Thus ΔH for the overall process is +1.0 kcal. For step (1), ΔS is +0.0543 (leads to more disorder), and for step (2), ΔS is -0.0440 (ions are more ordered in solution than in gas phase). The overall ΔS is +0.0103 kcal/°K. Substituting into the Gibbs-Helmholtz equation:

$$\Delta G = \Delta H - T\Delta S$$

$$\Delta G = +1.0 \text{ kcal} - (298^\circ K) \ (+0.0103 \text{ kcal/}^\circ K)$$

$$\Delta G = +1.0 \text{ kcal} - 3.07 \text{ kcal} = -2.07 \text{ kcal}$$

This is a classic example of the effect of the TΔS term. On the basis of ΔH alone, ΔG would be positive and NaCl would not dissolve at room temperature!

Example 5.

Based on data given in Example 4, calculate ΔG for the solution of NaCl at 0°C and 100°C at 1 atm pressure.

Solution.

When values of ΔH and ΔS (1 atm) are known, ΔG may be determined at any temperature by substitution into the Gibbs-Helmholtz equation. For the solution of NaCl, ΔH is +1.0 kcal and ΔS is +0.0103 kcal/°K. Thus, for 0°C (273°K) we find:

$$\Delta G = \Delta H - T\Delta S$$
$$= +1.0 \text{ kcal} - (273°\text{K}) (+0.0103 \text{ kcal/°K})$$
$$= +1.0 \text{ kcal} - 2.81 \text{ kcal}$$
$$= -1.81 \text{ kcal}$$

for 100°C (373°K),

$$\Delta G = +1.0 \text{ kcal} - (373°\text{K}) (+0.0103 \text{ kcal/°K})$$
$$= +1.0 \text{ kcal} - 3.84 \text{ kcal}$$
$$= -2.84 \text{ kcal}$$

Notice that ΔG becomes more negative with increasing temperature. This means, of course, that NaCl becomes more soluble with increasing temperature. This effect is due to the positive ΔS term.

When a substance undergoes a phase change at its melting or boiling point, ΔG for the process is zero, i.e., the change may occur spontaneously and simultaneously in either direction, solid $\rightleftarrows$ liquid or liquid $\rightleftarrows$ gas. So, when ΔH for the phase change is known, the entropy change for the process can be found from the Gibbs-Helmholtz equation.

Example 6.

At 1 atm pressure silicon tetrachloride melts at a temperature of −68°C and boils at a temperature of 57°C. For the phase change from solid to liquid, ΔH (called the heat of fusion) is 1840 cal/mole. For the phase change from liquid to gas, ΔH (called the heat of vaporization) is 6140 cal/mole. Find ΔS (1 atm) for

$$SiCl_4 (s) \rightarrow SiCl_4 (l) \text{ at } -68°C$$

and

$$SiCl_4 (l) \rightarrow SiCl_4 (g) \text{ at } 57°C$$

Solution.

Setting ΔG = 0 in the Gibbs-Helmholtz equation we have

$$\Delta G = \Delta H - T\Delta S$$
$$0 = \Delta H - T\Delta S$$

and

$$\Delta S = \frac{\Delta H}{T}$$

For $SiCl_4$ (s) → $SiCl_4$ (l) at −68°C (205°K),

$$\Delta S = \frac{1840 \text{ cal/mole}}{205°K} = 9.0 \text{ cal/mole} - °K$$

For $SiCl_4$ (l) → $SiCl_4$ (g) at 57°C (330°K),

$$\Delta S = \frac{6140 \text{ cal/mole}}{330°K} = 18.6 \text{ cal/mole} - °K$$

Note that melting or boiling a substance is an endothermic process and ΔH would be positive. Also, melting involves a breakdown of crystal structure that leads to more disorder in the liquid phase compared to the solid, and ΔS for a solid → liquid phase change is positive. A gas is more disordered than a liquid, and ΔS for a liquid → gas phase change is also positive. Conversely, ΔS for condensation (g → l) or freezing (l → s) is negative.

Category I

PROBLEMS

1. From Table 3–1 (bond energies), calculate ΔH (reaction) for
   (a) $N_2 + 3 H_2 \rightarrow 2 NH_3$
   (b) $H_2O + Cl_2 \rightarrow 2 HCl + \frac{1}{2} O_2$
   (c) $2 HI + Br_2 \rightarrow 2 HBr + I_2$

2. The ΔH (reaction) is −8.5 kcal for the reaction

$$NH_3 + 3 I_2 \rightarrow NI_3 + 3 HI$$

Calculate the bond energy of the N—I bond.

3. From Table 3–1 (bond energies), calculate ΔH (reaction) for
   (a) $I_2 + 2 HCl \rightarrow 2 HI + Cl_2$
   (b) Diamond + $2 Cl_2 \rightarrow CCl_4$
   (c) $H_2O + Br_2 \rightarrow 2 HBr + \frac{1}{2} O_2$

4. The heat of combustion of methane, CH_4, is listed at −211 kcal per mole. Calculate the C=O bond energy in carbon dioxide.

Category II

5. From Table 3–2 (heats of formation), calculate ΔH (reaction) for
   (a) $CO(g) + H_2O(g) \rightarrow CO_2(g) + H_2(g)$
   (b) $CaCO_3(s) + H_2SO_4(aq) \rightarrow CaSO_4(s) + H_2O(l) + CO_2(g)$
   (c) $H_2O_2(l) + 2 HI(g) \rightarrow I_2(s) + 2 H_2O(l)$

6. A very important organic chemistry reaction is the chlorination of ethylene to form 1,2-dichloroethane: $C_2H_4(g) + Cl_2(g) \rightarrow C_2H_4Cl_2(l)$. The heat of reaction is −52.2 kcal per mole. Find the heat of formation of $C_2H_4Cl_2(l)$.

7. From Table 3–2 (heats of formation), calculate ΔH (reaction) for
   (a) $SO_3(g) + H_2O(l) \rightarrow H_2SO_4(aq)$
   (b) $CaO(s) + SO_3(g) \rightarrow CaSO_4(s)$
   (c) $CaO(s) + H_2SO_4(aq) \rightarrow CaSO_4(s) + H_2O(l)$

8. Show how the three equations and the corresponding heats of reaction are all related in Problem 7.

9. Fortunately for the inhabitants of our planet, the following reaction is endothermic with a heat of reaction of +89.2 kcal:

$$CO_2(g) + \tfrac{1}{2} N_2(g) \rightarrow CO(g) + NO(g)$$

Calculate the heat of formation of $NO(g)$.

Category III

10. The heat of combustion of the gasoline constituent octane, C_8H_{18}, is -1300 kcal per mole. If 1.14 grams of octane are burned in a calorimeter containing 1000 g of water at $25.0°C$, what will be the temperature of the water after reaction?

11. An unknown acid is reacted with NaOH in a calorimeter containing 500 g of water. The water temperature rose from $26.22°C$ to a minimum of $28.54°C$. Calculate the heat of the reaction.

12. Using a known reaction, the heat capacity of a calorimeter is found to be 1.54 kcal per $°C$. When 1.04 g of chromium metal, Cr, is oxidized to $Cr_2O_3(s)$ in the calorimeter, the temperature increases from $25.0°C$ to $26.75°C$. Find the heat of formation of $Cr_2O_3(s)$ in kcal per mole.

Category IV

13. For the reaction $NO(g) \rightarrow \tfrac{1}{2} N_2(g) + \tfrac{1}{2} O_2(g)$, ΔH is -21.6 kcal and ΔS is -2.3 cal/$°K$ at $25°C$. Calculate ΔG for the reaction.

14. From the information in Problem 13, estimate the temperature at which the nitrogen and oxygen in our atmosphere would react spontaneously to produce NO.

15. Would you expect ΔS to be positive, negative, or near zero for the following:
    (a) water freezes
    (b) air is let out of a tire
    (c) a leaf forms on a tree
    (d) a crystal melts
    (e) a man dies
    (f) $H_2(g) + O_2(g) \rightarrow H_2O(g)$
    (g) the universe expands

16. From the table of free energies of formation, calculate ΔG (1 atm), $25°C$, for the following reactions:
    (a) $BaO(s) + CO_2(g) \rightarrow BaCO_3(s)$
    (b) $H_2S(g) + 2 O_2(g) \rightarrow SO_3(g) + H_2O(l)$
    (c) $NO(g) + \tfrac{1}{2} O_2(g) \rightarrow NO_2(g)$
    (d) $PbO_2(s) + CO(g) \rightarrow PbO(s) + CO_2(g)$
    (e) $H_2S(g) + Br_2(l) \rightarrow S°(s) + 2 HBr(g)$

17. From free energies of formation and heats of formation, calculate ΔS at $25°C$ and 1 atm for the reaction

$$CO(g) + \tfrac{1}{2} O_2(g) \rightarrow CO_2(g)$$

18. When silver bromide, AgBr, melts at $430°C$ and 1 atm pressure, ΔH for the process is 2180 calories per mole. Find ΔS for the melting of one mole of silver bromide.

19. The heat of vaporization of ethyl alcohol is 9400 calories per mole at its standard boiling point (78°C). Find ΔS for the vaporization of 2.30 grams of ethyl alcohol at 78°C and 1 atm pressure.

20. For the reaction $CO(g) + \frac{1}{2} O_2(g) \rightarrow CO_2(g)$, we found ΔS (1 atm) = -0.0208 kcal/mole-°K (Problem 17), and $\Delta H = -67.7$ kcal. Calculate ΔG (1 atm) for the reaction at
(a) 2000°K
(b) 4000°K

General Problems

21. Calculate the heats of reaction for the following from bond energies and from heats of formation. Compare the results. Explain your results for (b).
(a) $H_2O(g) + Cl_2(g) \rightarrow 2 HCl(g) + \frac{1}{2} O_2(g)$
(b) $2 HI(g) + \frac{1}{2} O_2(g) \rightarrow I_2(s) + H_2O(g)$

22. Exactly 4.345 g of MnO_2 were produced by oxidation of Mn metal in a calorimeter. The calorimeter contained 1000 g of water, and during the reaction the temperature rose from 26.23°C to 32.46°C. The heat of reaction for $MnO(s) + \frac{1}{2} O_2(s) \rightarrow MnO_2(s)$ is -32.5 kcal/mole. Find the heat of formation for $MnO(s)$.

23. Find ΔH of reaction (1 atm, 25°C) for the following:
(a) Exactly 10.0 g of calcium carbonate decomposes:

$$CaCO_3(s) \rightarrow CaO(s) + CO_2(g)$$

(b) Exactly 12.6 g of liquid HNO_3 is dissolved in water:

$$HNO_3(l) \rightarrow HNO_3(aq)$$

24. Find ΔH, ΔG, and ΔS at 25°C and 1 atm for the reaction

$$CaO(s) + SO_3(g) \rightarrow CaSO_4(s)$$

4 GASES

Studies of the behavior of gases were historically fundamental to the birth of chemistry as a science, and continue, today, to be the basis for many important developments in theoretical chemistry. Atoms, molecules, or ions are compacted in the liquid or solid phase, but widely dispersed in the gas phase. One mole of a substance as a liquid or solid occupies a volume of approximately 20 ml (e.g., 18 ml for liquid H_2O), whereas one mole of a gas at STP (1.0 atm pressure and 0°C) occupies a volume of 22.4 liters. Thus, the same number of particles occupies a thousandfold larger volume as a gas than as a liquid or solid. Consequently, molecules in the gas phase may be treated as behaving (almost) independently, without regard to attractive or repulsive forces that are important in the solid or liquid phase.

An understanding of the behavior of gases requires, first, an understanding of the terms volume, pressure, and temperature.

Volume is easily visualized as a certain number of cubic centimeters, cubic inches, liters, quarts, etc. A gas always fills the container in which it is placed, and the volume of a gas is simply the volume of the vessel enclosing the gas.

Pressure is a measure of the direction of mass flow. Mass will move from a higher pressure to a lower pressure (e.g., water out of a tap). Pressure is a force per unit area, such as pounds per square inch or grams per square centimeter. Gas pressures are usually measured in atmospheres (atm) or Torr (mm Hg). The latter units may be understood by examining "atmosphere pressure." The 500-mile thick blanket of air around the earth exerts a pressure of 14.7 lbs per square inch on objects at sea level. If one inverts a tube full of liquid mercury into a dish of mercury, not all of the mercury will run out of the tube into the dish. Regardless of the diameter of the tube, a column of mercury approximately 760 mm high will remain in the tube (at sea level). That column of mercury is supported by the pressure of the atmosphere pushing on the surface of the mercury in the dish. Thus, 1.0 atmosphere of pressure is defined at 760 mm Hg (or 760 Torr).

Temperature is a measure of the direction of heat flow. Heat energy will move from a higher temperature to a lower temperature. The most common units of temperature (in science) are degrees Celsius or centigrade. On the Celsius scale, 0°C is defined as the freezing temperature of water at 1.0 atm pressure and 100°C as the boiling temperature of water at 1.0 atm. There is a lower limit to temperature, where all molecular or atomic motion ceases, and this temperature is −273.15°C or *absolute zero*. The Kelvin temperature scale defines 0°K as this lower limit. The relationship between the Celsius and Kelvin scale is

$$°C + 273.15 = °K$$

Gas Laws

Thousands of experiments conducted in the 17th, 18th, and 19th centuries have resulted in what are known as "gas laws." Gas laws are concise statements that describe the physical behavior of most gases. They are

Boyle's Law: For a fixed amount of gas at constant temperature, the pressure of the gas is inversely proportional to its volume, $P \propto 1/V$.

Charles' Law: For a fixed amount of gas at constant pressure, the volume of the gas is directly proportional to the absolute temperature, $V \propto T$.

Avogadro's Principle: At a constant temperature and pressure, the volume of any gas is directly proportional to the number of moles of the gas, $V \propto n$.

The Ideal Gas Law

Boyle's Law, Charles' Law, and Avogadro's Principle may be combined. Taking the proportionalities $P \propto 1/V$, $V \propto T$, and $V \propto n$, and inserting a proportionality constant, R, we get an equation: $PV = nRT$. This expression is called the *Ideal Gas Law.*

Dalton's Law: The total pressure of a mixture of gases is the sum of the pressures of each gas if it were alone in the container: $P(total) = P_1 + P_2 + P_3 \ldots$ etc.

Graham's Law: The rate of diffusion (or effusion) of a gas is inversely proportional to the square root of the molecular weight of the gas:

$$\text{Rate} \propto 1/\sqrt{MW}$$

Most gas problems in chemistry may be solved by suitable application of the Ideal Gas Law. The problems in this chapter are divided into three categories.

Problem Categories

I Various uses of $PV = nRT$.
II Calculations involving initial and final states.
III Uses of Dalton's and Graham's Laws.

USING THE IDEAL GAS LAW   CATEGORY

I

The proportionality constant, R, (called the ideal gas constant) can have different values depending upon the units of P, V, and T. At 1.0 atm pressure and $0°C$ ($273°K$), one mole of any ideal gas occupies a volume of 22.4 liters. Based upon these units, the value of R is

$$R = \frac{PV}{nT} = \frac{(1.0 \text{ atm}) (22.4 \text{ } \ell)}{(1.0 \text{ mole}) (273°K)} = 0.0821 \frac{\text{atm} - \ell}{\text{mole} - °K}$$

If, for example, the units of pressure were in mm Hg rather than atmospheres, then the value of R would be 62.4 $\frac{mm\ Hg - \ell}{mole - °K}$. What would be the value of R if pressure were in mm Hg, volume in milliliters, and temperature in °K?

Example 1.

What volume would 14.2 g of chlorine gas, Cl_2, occupy at a pressure of 380 mm and a temperature of 25°C?

Solution.

Gas law problems usually contain several known quantities and only one unknown. Problems of this type cannot be solved in a single set-up using the factor-label method. The best approach is simply to "plug into" the Ideal Gas Law. The main thing to watch out for is that the units you *use* match the units you *choose* for the ideal gas constant. Let's use 0.0821 $\frac{atm. - \ell}{mole - °K}$ as our value of R in this problem. Then, P must be in atm:

$$380\ mm\ Hg \times \frac{1\ atm}{760\ mm\ Hg} = 0.500\ atm$$

temperature must be in °K:

$$25°C\ is\ 273 + 25 = 298°K$$

and grams of Cl_2 must be converted to moles:

$$14.2\ g\ Cl_2\ are\ 14.2/70.9 = 0.200\ mole\ Cl_2$$

and we can solve for the volume in liters:

$$PV = nRT;\quad V = \frac{nRT}{P}$$

$$V = \frac{(0.200\ mole)\ (0.0821\ atm\text{-}\ell/mole\text{-}°K)\ (298°K)}{(0.50\ atm)} = 9.79\ liters$$

If we use 62.4 $\frac{mm\ Hg\text{-}\ell}{mole\text{-}°K}$ as our value of R, then we can substitute P into the equation in mm Hg, without converting to atmospheres of pressure.

$$V = \frac{(0.200\ mole)\ (62.4\ mm\ Hg\text{-}\ell/mole\text{-}°K)\ (298°K)}{(380\ mm\ Hg)} = 9.79\ liters$$

Example 2.

In a laboratory experiment a student places an unknown liquid in a 500.0 ml flask equipped with a very small outlet. The liquid is vaporized at 80.0°C and 740.0 mm pressure until all excess of the vapor is expelled and the flask is just filled with the unknown vapor. The weight of the vapor was found to be 0.556 gram. Calculate the molecular weight of the unknown.

Solution.

One of the many applications of the Ideal Gas Law is the determination of molecular weights of gases or substances that are easily vaporized. The number of moles, n, of a molecular substance is wt/MW; hence, if P, V, T, and the wt are known, one can easily find the MW. Using R = 0.0821 ℓ-atm/mole-°K, we can "plug in"; PV = (wt/MW)RT,

$$\left(\frac{740}{760} \text{ atm}\right)(0.500 \text{ ℓ}) = \left(\frac{0.556 \text{ g}}{MW}\right)(0.0821 \text{ ℓ-atm/mole-°K})(353°K),$$

$$MW = \frac{0.556 \times 0.0821 \times 353}{0.974 \times 0.500} = 33.1 \text{ g/mole}$$

Example 3.

The density of an unknown gas is 1.12 grams per liter at 25.0°C and 640 mm pressure. Find the molecular weight of the gas.

Solution.

Writing the Ideal Gas Law in the form

$$PV = \frac{wt}{MW} \times R \times T$$

we can rearrange to give

$$P \times MW = \left(\frac{wt}{V}\right) \times R \times T$$

The term $\left(\dfrac{wt}{V}\right)$ is simply density in grams per liter. Thus, at a given P and T, molecular weight can be determined from gas density, or vice-versa. In this example

$$MW = \frac{1.12 \text{ g/ℓ} \times 0.0821 \text{ ℓ-atm/mole-°K} \times 298°K}{0.842 \text{ atm}} = 32.5 \text{ g/mole}$$

Example 4.

Aluminum metal dissolves in strong base by the reaction

$$2 \text{ Al(s)} + 2 \text{ OH}^-(aq) + 6 \text{ H}_2O \rightarrow 2 \text{ Al(OH)}_4^-(aq) + 3 \text{ H}_2 \text{ (g)}$$

At 700 mm Hg pressure and 30.0°C, what volume of hydrogen gas could be produced from 5.00 grams of aluminum?

Solution.

First, find the number of moles of H_2 gas produced. Remember those stoichiometry problems?

$$X \text{ moles H}_2 = 5.00 \text{ g Al} \times \frac{1 \text{ mole Al}}{27.0 \text{ g Al}} \times \frac{3 \text{ moles H}_2}{2 \text{ moles Al}} = 0.278 \text{ mole H}_2$$

Then substitute into the Ideal Gas Law.

$$P \times V = n \times R \times T$$

$$\left(\frac{700}{760} \text{ atm}\right) \times V = 0.278 \text{ mole} \times 0.0821 \text{ atm-}\ell\text{/mole-}^\circ K \times 303^\circ K$$

$$V = 7.51 \text{ liters}$$

Example 5.

If 2.66 grams of a "freon" gas, $C_2Cl_2F_2$, are placed in a 250 ml hair spray container, what will be the pressure in the container at $-70^\circ C$, $25^\circ C$, and $100^\circ C$?

Solution.

First,

$$X \text{ moles } C_2Cl_2F_2 = 2.66 \text{ g } C_2Cl_2F_2 \times \frac{1 \text{ mole } C_2Cl_2F_2}{133.0 \text{ g } C_2Cl_2F_2} = 0.020 \text{ mole}$$

Then, substituting into $PV = nRT$, and solving for P, we find

$$P(-70^\circ C) = \frac{0.020 \text{ mole} \times 0.0821 \text{ }\ell\text{-atm/mole-}^\circ K \times 203^\circ K}{0.250 \text{ liter}} = 1.33 \text{ atm}$$

$$P(25^\circ C) = \frac{0.020 \times 0.0821 \times 298}{0.250} = 1.96 \text{ atm}$$

$$P(100^\circ C) = \frac{0.020 \times 0.0821 \times 373}{0.250} = 2.45 \text{ atm}$$

If the container explodes under 3.0 atm pressure, at what temperature will explosion occur?

CATEGORY II   **CALCULATIONS INVOLVING INITIAL AND FINAL STATES**

Quite often gas problems involve a change in P, V, n, or T from an initial to a final set of conditions. When that is the case the desired relationships may be obtained from the Ideal Gas Law.

Example 1.

In the first example in Category I, it was found that 14.2 g of Cl_2 gas occupies a volume of 9.79 liters at 380 mm Hg and $25^\circ C$. What volume will the Cl_2 occupy at 1040 mm Hg and $100^\circ C$?

Solution.

Knowing the amount of Cl_2, we could solve the problem by plugging into $PV = nRT$ for the new P and T, or we can proceed as follows. We know that, for the first set of conditions, the Ideal Gas Law will hold true and that

$$P_1V_1 = nRT_1$$

We know that the Law will also apply to the second set of conditions, and that

$$P_2V_2 = nRT_2$$

Now, since both n and R remain constant for both sets of conditions, we can set:

$$\frac{P_1 V_1}{T_1} = \frac{P_2 V_2}{T_2}$$

In this particular problem we wish to find the new volume, V_2, and we can rearrange the equation to give

$$V_2 = V_1 \times \frac{P_1}{P_2} \times \frac{T_2}{T_1}$$

Thus, we can say the final volume, V_2, is equal to the initial volume, V_1, times a pressure change factor and a temperature change factor. For our problem,

$$V_2 = 9.79 \, \ell \times \frac{380 \text{ mm Hg}}{1040 \text{ mm Hg}} \times \frac{373°\text{K}}{298°\text{K}} = 4.48 \, \ell$$

Note that while T must be in °K, P could be used either as mm Hg or as atm. Why? Also, notice that "common sense" tells us how the pressure change and temperature change factors must appear without having to remember P_1/P_2 or T_2/T_1. In our example, the pressure is increased from 380 to 1040 mm. What effect does an increase in pressure have on the volume of a gas? It will tend to decrease the volume, so it is logical that we multiply V_1 by a factor less than one, 380/1040. The temperature increases from 298°K to 373°K. What effect will an increase in temperature have on the volume of a gas? Common sense dictates that the temperature increase would tend to increase the volume and that we would multiply V_1 by 373/298, a factor greater than 1.0.

Example 2.

The gas in a weather balloon occupies a volume of 2000 liters at sea level (P = 760 mm Hg) on a warm day (30°C). When the balloon ascends to 25,000 ft, the temperature and pressure (both inside and outside the balloon) drop to −25°C and 250 mm Hg. What is the volume of the gas in the balloon at 25,000 ft?

Solution.

$$V_2 = V_1 \times \frac{P_1}{P_2} \times \frac{T_2}{T_1} = 2000 \, \ell \times \frac{760 \text{ mm}}{250 \text{ mm}} \times \frac{248°\text{K}}{303°\text{K}} = 5000 \, \ell$$

It is easy to see why balloons are only partially inflated at "lift off."

Example 3.

A gas in a closed steel container exerts a pressure of 1.00 atm at 0°C. What will the pressure be at 500°C?

Solution.

The expression,

$$\frac{P_1 V_1}{T_1} = \frac{P_2 V_2}{T_2}$$

can be rearranged to solve for any of the various parameters in the equation. In this case, we want a new pressure, P_2, and we can write

$$P_2 = P_1 \times \frac{T_2}{T_1} \times \frac{V_1}{V_2}$$

We can assume that the volume of the steel container remains constant, hence the V_1/V_2 term is just equal to 1.0. Then,

$$P_2 = P_1 \times \frac{T_2}{T_1} = 1.00 \text{ atm} \times \frac{773°K}{273°K} = 2.83 \text{ atm}$$

CATEGORY III   DALTON'S AND GRAHAM'S LAWS

Dalton's Law of Partial Pressures and Graham's Law of Diffusion (or Effusion) are of particular importance when dealing with mixtures of gases. Their uses are best illustrated by example.

Example 1.

An "atmosphere" inside a space capsule uses a mixture of oxygen and helium. If the volume of the capsule is 15,000 liters, the temperature is to be maintained at 20°C, and a mixture of 600 grams of He and 5000 grams of O_2 are released in the capsule, what is the "atmospheric" pressure inside the capsule?

Solution.

Dalton's Law states that the total pressure of a gas mixture will be the sum of the partial pressures of each of the gases in the mixture. The partial pressure is the pressure a gas would have if it were by itself.

Using the Ideal Gas Law, we can find the pressure due to the helium:

$$P_{He} = \left(\frac{wt}{MW}\right) \times \frac{RT}{V} = \frac{600 \text{ g} \times 0.0821 \text{ }\ell\text{-atm/mole-}°K \times 293°K}{4.0 \text{ g} \times 15,000 \text{ }\ell} = 0.24 \text{ atm}$$

and the oxygen,

$$P_{O_2} = \left(\frac{wt}{MW}\right) \times \frac{RT}{V} = \frac{5000 \text{ g} \times 0.0821 \text{ }\ell\text{-atm/mole-}°K \times 293°K}{32.0 \text{ g} \times 15,000 \text{ }\ell} = 0.25 \text{ atm}$$

By Dalton's Law:

$$P_{total} = P_{He} + P_{O_2} = 0.24 \text{ atm} + 0.25 \text{ atm} = 0.49 \text{ atm}$$

If you were an astronaut, would that be acceptable?

Example 2.

In the laboratory, gases are often collected by displacing water in a flask. When a gas is trapped in this manner, the flask always contains some water vapor. If 1500 ml of O_2 were collected over water at 25°C and 720 mm pressure, how many grams of O_2 and how many grams of H_2O vapor would there be in that 1500 ml? At 25°C the vapor pressure (i.e., the partial pressure) of water vapor is 24 mm Hg.

Solution.

From Dalton's Law, the total pressure is 720 mm Hg = $P_{O_2} + P_{H_2O}$. Since P_{H_2O} is 24 mm Hg, P_{O_2} = 696 mm Hg. We can solve the Ideal Gas Law for wt (O_2) and wt (H_2O).

$$PV = \left(\frac{wt}{MW}\right) RT; \quad wt = \frac{P \times V \times MW}{R \times T}$$

$$Wt\ (O_2) = \frac{696\ mm\ Hg \times 1.50\ \ell \times 32.0\ g}{62.4\ mm\ Hg\text{-}\ell/mole\text{-}°K \times 298°K} = 1.80\ g$$

$$Wt\ (H_2O) = \frac{24\ mm\ Hg \times 1.50\ \ell \times 18.0\ g}{62.4\ mm\ Hg\text{-}\ell/mole\text{-}°K \times 298°K} = 0.035\ g$$

Note the units used for P and R.

Example 3.

A mixture of He and O_2 are injected into one end of an evacuated tube. Which gas will reach the other end of the tube first? What are the relative rates of effusion of the two gases?

Solution.

Graham's Law states that the rate of effusion (movement through vacuum) or diffusion (movement through another gas) is inversely proportional to $\sqrt{MW}$ of a gas. Thus, the lower the MW (or atomic wt for He), the faster the rate of effusion, and He would reach the other end of the tube first.

Graham's Law may be stated as an equation if a proportionality constant is used:

$$Rate = C/\sqrt{MW}$$

For two different gases under the same conditions of temperature and pressure, C is the same for both and we can set

$$Rate_1 \times \sqrt{MW_1} = Rate_2 \times \sqrt{MW_2}$$

If we call He gas 1 and O_2 gas 2, then

$$Rate\ (He) = Rate\ (O_2) \times \frac{\sqrt{32.0}}{\sqrt{4.0}} = Rate\ (O_2) \times 2.8$$

Thus, He effuses 2.8 times faster than O_2.

In this form, Graham's Law may be used to estimate the molecular weight of an unknown gas by comparing its rate of effusion with that of a known gas. How?

Example 4.

Samples of "laughing gas," N_2O, and a "tear gas," $C_6H_{11}OBr$, are simultaneously released at the front of a classroom. If students in the front row begin to laugh after 30 seconds, how long will it be before they begin to cry?

Solution.

The relative rates of diffusion (see Example 3) are

$$Rate\ (N_2O) = Rate\ (C_6H_{11}OBr) \times \frac{\sqrt{C_6H_{11}OBr}}{\sqrt{N_2O}} = Rate\ (C_6H_{11}OBr) \times \frac{\sqrt{179}}{\sqrt{44}}$$

The rate (N_2O) is 2.0 times the rate $(C_6H_{11}OBr)$, so, if it takes 30 seconds for the laughing gas to reach the front row, it will take 60 seconds for the tear gas to get there.

PROBLEMS   *Category I*

1. What volume will be occupied by 2.20 grams of carbon dioxide at STP?

2. If 0.20 mole of an ideal gas occupies a volume of 8.96 liters at 1.00 atm pressure, what is the temperature of the gas – in °K and °C?

3. What is the volume of 22.4 grams of CO at 700 mm pressure and 27.0°C?

4. Data from the Viking Landers on the surface of Mars indicate an atmosphere that is 95% CO_2, with traces of N_2, Ar, O_2, CO, and O. The atmospheric pressure and temperature of Mars varies, but typical values are 7.6 millibars pressure and 250°K. Calculate the molar concentration (moles per liter) of the Martian atmosphere. (1 bar = 0.987 atm)

5. Probes of the atmosphere of the planet Venus indicate an atmosphere of CO_2 at a pressure of 50 atmospheres and a temperature of 700°K. Find the density (g/ℓ) of the atmosphere of Venus for those particular conditions.

6. A sample of He gas in a sealed container has a density of 1.50 g/ℓ at 27°C. What is the pressure in the container?

7. A volatile liquid was vaporized in a 250.0 ml flask at 90°C and 740 mm pressure. The condensed vapor weighed 0.722 grams. Find the molecular weight of the liquid.

8. If 10.83 grams of mercuric oxide, HgO, is completely decomposed into Hg° and $O_2(g)$, what volume of O_2 will be produced at 60°C and 680 mm pressure?

Category II

9. A blimp was filled with He to a volume of 40,000 liters at sea level (p = 760 mm) when the temperature was 30°C. When the blimp was high above the Super Bowl, the weather changed. The pressure dropped to 700 mm and the temperature to –3.0°C. Would the blimp expand or contract? By how much?

10. A gas in a sealed container exerted a pressure of 1.20 atm at 20°C. What would be the pressure at 200°C?

11. Five liters of hydrogen gas were collected at 25°C and 710 mm. It was then compressed by putting it under 2.20 atm pressure at a temperature of –30°C. Find the volume of the compressed hydrogen.

12. The gas in a hot air balloon has a volume of 6000 liters at 50°C. To what temperature must the gas be raised in order to increase the volume of the balloon to 7000 liters?

Category III

13. Use the data and results from Problem 4 to calculate the partial pressure of O_2 in the Martian atmosphere if O_2 is 1.0% (by volume) of the total atmosphere of Mars.

14. The contents of a scuba diver's tank are 20% O_2, 78% N_2 and 2% Ar by volume. The total pressure within the tank is 1.20 atm at 10°C. Find the partial pressure of O_2 in the tank.

15. An engineer testing the Alaskan pipeline releases large amounts of CO_2 and C_2F_6 into the pipeline at Point Barrow. Exactly 26.0 hours later, the CO_2 is detected 500 miles down the pipe. Assuming no leaks, how long a time will it take for the C_2F_6 to reach the 500-mile mark?

16. A burst of solar energy from the sun releases millions of tons of He and other gases. Assume a solar flare was observed on earth on January 3 at 3:10 p.m. The first He from the flare reached earth's atmosphere at 5:30 p.m. on January 4. At 9:30 a.m. on January 5, an unknown gas entered the earth's atmosphere. Estimate the "molecular" weight of the unknown gas.

General Problems

17. Find the density, in g/ℓ, of gaseous radon, Rn, at 10°C and 1 atm pressure.

18. If 15.0 g of CO gas occupy a volume of 18.0 liters at a certain temperature and pressure, what volume in liters will 30.0 g of C_2H_4 gas occupy at the same temperature and pressure?

19. Under pressure of 44.8 atm, at what temperature will He gas have a density of 8.0 g/ℓ?

20. An unknown gas diffuses through a length of glass tubing at a rate of 4 cm per second. Under the same conditions, CH_4 diffuses through the tubing at 6 cm per second. Find the approximate molecular weight of the unknown gas.

21. In a gaseous mixture of F_2 and Cl_2 there are twice as many moles of F_2 as Cl_2. The gases are confined in a cylinder. The pressure due to Cl_2 is 640 mm Hg. What is total pressure in the cylinder?

22. What volume will be occupied by a mixture of 10.0 g of argon and 8.80 g of carbon dioxide at a pressure of 288 mm Hg and a temperature of 47°C?

23. Acetylene gas, used for welding, is at a pressure of 5.0 atm in a 20.0 liter cylinder at 25°C. What will the pressure in the cylinder be at 150°C?

24. To what temperature must 0.720 g of water vapor be heated in a 500 ml flask before the pressure in the flask becomes 2.0 atm?

25. A 6.00 liter gas sample at 0.250 atm and 100°K is heated to 400°K and the pressure is increased to 0.500 atm. What volume will the gas occupy at 400°K and 0.500 atm?

5 ATOMIC STRUCTURE AND BONDING

Subjects included under this heading may vary considerably both in scope and complexity, ranging from an elementary explanation of subatomic particles to the most sophisticated mathematics of quantum theory. In most introductory chemistry courses, atomic structure and chemical bonding are treated as non-mathematical subjects (to the delight and relief of the students!). Consequently, there are few, if any, problems requiring mathematics. However, some subject areas do pose "problems" that may not require computation, but do need explanation and practice. Let us examine a few of these. Detailed explanations of orbitals, subshells, etc., must be left to the textbooks, but the problems presented here should provide you with some practice and helpful hints.

Problem Categories

    I Writing electron configurations.
   II Drawing dot (Lewis) formulas.
  III Determining shapes of molecules.

CATEGORY I WRITING ELECTRON CONFIGURATIONS

The ability to write electron configurations for neutral atoms and for positive and negative ions is fundamental to an understanding of chemistry. One simple and useful device to remember is the "diagonal arrow" diagram (Fig. 5–1). If one

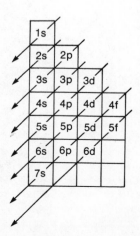

stacks up the various subshells as they occur in each main period of elements, the *order of occupancy* by electrons follows a diagonal pattern. The diagonal arrows indicate that the order in which the subshells are occupied by electrons is first, the 1s, then the 2s, then the 2p followed by the 3s, then the 3p followed by the 4s, then the 3d, 4p, and 5s, and so on. One need only remember that any s subshell can accommodate up to 2 electrons, any p subshell up to 6 electrons, and any d subshell up to 10, and any f subshell up to a total of 14 electrons. The atomic number of an element gives the total number of electrons on the atom, so writing the electron configuration of an element becomes a simple matter of placing the correct total number of electrons in the appropriate subshells.

Example 1.

Write the electron configuration for the oxygen atom.

Solution.

Oxygen is element number 8, so we must account for a total of 8 electrons. The diagram tells us the 1s subshell is occupied first, so the first two electrons go there, then the next two go in the 2s, then we go to the 2p subshell. A p subshell can hold up to 6 electrons, but if we filled it, that would give a total of 10 electrons and we need only 8, so the 2p must contain only 4 electrons. By using a horizontal notation, we can write the electron configuration for oxygen as

$$O \qquad 1s^2 2s^2 2p^4$$

Example 2.

Write the electron configuration for the iron (Fe) atom.

Solution.

We need a total of 26 electrons. We can begin: $1s^2 2s^2 2p^6 3s^2 3p^6$ .... At this point, we have 18 electrons and need 8 more. From the order in which the subshells are *occupied*, we see that the "next" two of those 8 electrons will be placed in the 4s subshell and we can write $4s^2$. We still have 6 electrons to be accounted for. Our "diagonal arrow" shows us that the 3d subshell is occupied next, and since a d subshell can hold up to 10 electrons, we can write out the complete configuration:

$$Fe \qquad 1s^2 2s^2 2p^6 3s^2 3p^6 4s^2 3d^6$$

or

$$Fe \qquad 1s^2 2s^2 2p^6 3s^2 3p^6 3d^6 4s^2$$

Either way is correct, but the latter sequence, where the subshells of a principal quantum level (period) are grouped together has the advantage of indicating the order of *increasing energy* of the subshells as one reads from left to right. Although we say "the 4s level is occupied before the 3d," as indeed it is for elements 19 and 20, it is important to note that beyond element number 20 (calcium), the 4s subshell is actually higher in energy than the 3d subshell. This fact is particularly significant when deciding which electrons "come off" in the formation of positive ions. See Example 3.

Example 3.

Write the electron configuration of Ni^{2+} ion.

Solution.

Following our rules, we can write out the configuration for elemental nickel, Ni°, as

$$Ni \qquad 1s^2\,2s^2\,2p^6\,3s^2\,3p^6\,4s^2\,3d^8$$

or

$$Ni \qquad 1s^2\,2s^2\,2p^6\,3s^2\,3p^6\,3d^8\,4s^2$$

If we now remove 2 electrons from Ni° to make Ni^{2+}, which electrons come off? The electrons in the highest energy level will be the most easily removed and the highest energy level on a nickel atom is the 4s subshell. So, the electron configuration of Ni^{2+} is

$$Ni^{2+} \qquad 1s^2\,2s^2\,2p^6\,3s^2\,3p^6\,3d^8$$

It is *not*

$$1s^2\,2s^2\,2p^6\,3s^2\,3p^6\,4s^2\,3d^6$$

Example 4.

Write the electron configurations for tin (Sn) and lead (Pb) atoms.

Solution.

Follow the diagonal arrow diagram to give

$$Sn \qquad 1s^2\,2s^2\,2p^6\,3s^2\,3p^6\,4s^2\,3d^{10}\,4p^6\,5s^2\,4d^{10}\,5p^2$$

Rearranging so the subshells of each principal level are grouped together, we get the configuration by order of increasing energy:

$$Sn \qquad 1s^2\,2s^2\,2p^6\,3s^2\,3p^6\,3d^{10}\,4s^2\,4p^6\,4d^{10}\,5s^2\,5p^2$$

It gets tedious writing out all the subshells for the heavier elements and often a short-cut notation is used which indicates only those electrons lying above the previous noble gas. For example, the "last" noble gas in the periodic table before tin is krypton, number 36. We know that Kr contains all the electrons, $1s^2 \ldots 4p^6$, so for tin we can write simply

$$Sn \qquad (Kr\ core)\ 4d^{10}\,5s^2\,5p^2$$

The electron configuration for lead, Pb, is similar, but don't forget to include the 4f subshell. The configuration may be written:

$$Pb \qquad (Xe\ core)\ 4f^{14}\,5d^{10}\,6s^2\,6p^2$$

Write the configurations for Sn^{2+} and Sn^{4+}.

Example 5.

Write the electron configuration for a sulfide ion, S^{2-}.

Solution.

For negative ions, simply add the "extra" electrons where they would normally go if you were building up a heavier element. Thus, the configuration of S° is

$$S \quad (Ne\ core)\ 3s^2\ 3p^4$$

Since the 3p subshell can accommodate 6 electrons, the two extra electrons on S^{2-} will occupy that subshell, and the configuration is

$$S^{2-} \quad (Ne\ core)\ 3s^2\ 3p^6$$

Using the Periodic Table to Obtain Electron Configurations

With experience you will find that the easiest way to determine electron configurations is simply to look at the periodic table (the chemist's bible), and count! Examine a complete periodic table. Elements on the left side, groups IA and IIA, are those in which an s subshell is being occupied (Fig. 5–2). The 1s subshell is filled after H and He. Lithium, Li, has a single electron in the 2s subshell. Beryllium contains 2 electrons in the 2s subshell, at which point it is filled. At Na and Mg the 3s subshell is occupied, the 4s at K and Ca, and so on. Elements on the right side of the periodic table, groups IIIA through VIIIA, are those in which p subshells are being occupied. From B through Ne the 2p subshell is filled, from Al through Ar the 3p is filled, etc. In the middle, or "hollow," part of the periodic table, the 3d, 4d, and 5d subshells are filling, and those elements, written in horizontal columns at the bottom and called the Lanthanum Series and Actinium Series, represent the filling of the 4f and 5f subshells, respectively.

With a little practice, you can simply "read" the electron configuration of an element from the periodic table. Let's take phosphorus, P, for example. We know that at He the 1s was full ($1s^2$), at Be the 2s was full ($2s^2$), at Ne the 2p was full ($2p^6$), at Mg the 3s was full ($3s^2$), and phosphorus is in the region of the periodic table where the 3p subshell is filling. We see that Al would have one 3p electron, Si would have two, and P must have three. So, the configuration is

$$P \quad 1s^2\ 2s^2\ 2p^6\ 3s^2\ 3p^3$$

CATEGORY   **DRAWING DOT (LEWIS) FORMULAS**
II

Dot (or Lewis) formulas are a simple means of depicting chemical bonding in compounds of the A group elements. Dot formulas are seldom used with the B group elements. Dot formulas are based on the "octet rule," or the tendency of the A group elements to acquire noble gas ($s^2 p^6$) electron configurations either by losing, gaining, or sharing electrons. The valence electrons of an A group element are all the s and p electrons overlying the previous noble gas. Nitrogen, N, for example, has 5 valence electrons since there are 5 electrons ($2s^2 2p^3$) overlying the He configuration ($1s^2$). The group number of the A group elements signifies the number of valence electrons. Thus, any element in group IIA (Be, Mg, Ca, Sr, Ba, Ra) has two valence electrons, any element in group IIIA has three valence electrons, and so on. Dot formulas for the elements are simply the symbol of the element surrounded by dots representing the valence electrons. For example:

$$Na\cdot \qquad \cdot \overset{\cdot}{C}\cdot \qquad \cdot \overset{\cdot\cdot}{\underset{\cdot\cdot}{O}}\colon$$

Dot formulas for compounds depict how the elements attain noble gas structures, either by losing or gaining in ionic compounds;

$$Na\cdot + \cdot \overset{\cdot\cdot}{\underset{\cdot\cdot}{Cl}}\colon \rightarrow Na^+ \; \colon\!\overset{\cdot\cdot}{\underset{\cdot\cdot}{Cl}}\colon^-$$

or by sharing in covalent compounds:

$$\colon\!\overset{\cdot\cdot}{F}\cdot + \cdot\overset{\cdot\cdot}{F}\colon \qquad \rightarrow \colon\!\overset{\cdot\cdot}{F}\!\colon\!\overset{\cdot\cdot}{F}\colon$$

$$\cdot\overset{\cdot}{N}\cdot + \cdot\overset{\cdot}{N}\colon \qquad \rightarrow \colon\!N\colon\colon\colon N\colon$$

$$\cdot\overset{\cdot\cdot}{O}\colon + \cdot\overset{\cdot}{C}\cdot + \cdot\overset{\cdot\cdot}{O}\colon \rightarrow \colon\!\overset{\cdot\cdot}{O}\!\colon\colon\!C\!\colon\colon\!\overset{\cdot\cdot}{O}\colon$$

Binary (two element) *ionic* compounds are readily formed from elements of groups IA and IIA with those of groups VIA and VIIA, since the former can easily attain noble gas structures by losing one and two electrons, respectively, and the latter can obtain an "octet" by gaining two or one electron(s). Thus $\cdot Mg\cdot$ loses two electrons to become Mg^{2+}, while $\cdot\overset{\cdot\cdot}{O}\colon$ tends to gain two to become $\colon\!\overset{\cdot\cdot}{\underset{\cdot\cdot}{O}}\colon^{2-}$; $\colon\!\overset{\cdot\cdot}{\underset{\cdot\cdot}{Cl}}\colon$ can gain one electron to become $\colon\!\overset{\cdot\cdot}{\underset{\cdot\cdot}{Cl}}\colon^-$.

Most of the A group elements form covalent compounds in which valence electrons are shared. When drawing dot formulas for covalent compounds, remember two rules: (1) be sure you have the correct number of dots, corresponding to the total number of valence electrons of all atoms in the compound, and (2) where possible, surround each atom by eight dots (H being the exception).

Example 1.

Draw a dot formula for the ionic compound, calcium oxide, CaO.

Solution.

Calcium will attain an argon structure by losing electrons:

$$\cdot Ca\cdot \rightarrow Ca^{2+} \; (+ 2\, e^{-1}s)$$

and oxygen will complete its octet by gaining two electrons:

$$(2\,e^{-1}s\,+\,)\;\cdot\ddot{O}{:}\;\rightarrow\;{:}\ddot{O}{:}^{2-}$$

Thus, the dot formula of ionic CaO is:

$$Ca^{2+}\;{:}\ddot{O}{:}^{2-}$$

Example 2.

Predict the formula of the ionic compound formed between sodium and sulfur.

Solution.

Drawing dot formulas offers a convenient means of determining formulas of ionic (and covalent) compounds. In this case, sodium will tend to lose its single valence electron, $Na\cdot \rightarrow Na^+$ ($+1\,e^-$), while sulfur will gain two electrons:

$$\cdot\ddot{S}{:}\;(\;+2\,e^{-1}s)\;\rightarrow\;{:}\ddot{S}{:}^{2-}$$

In order to balance the gain and loss of electrons and combine Na^+ with S^{2-} to give a neutral compound, there must be two Na^+ for each S^{2-} and the formula is

$$Na_2S$$

Example 3.

The bonding in H_2O is covalent. Draw a dot formula for the H_2O molecule.

Solution.

Oxygen has six valence electrons. A hydrogen atom has only one electron. Hydrogen cannot possibly complete an "octet," but it can attain a noble gas (He) structure by sharing its one electron with another atom. Thus, for H_2O, we can write a dot formula:

$$H\cdot\;+\;{:}\ddot{O}{:}\;+\;H\cdot\;\rightarrow\quad \begin{array}{c} {:}\ddot{O}\cdot \\ H\ \ H \end{array}$$

The oxygen atom shares two of its valence electrons, one with each hydrogen atom. The oxygen atom thereby completes its octet and each H atom acquires a noble gas structure. Each covalent O—H bond is called a single bond since only one *pair* of electrons is shared between atoms.

Example 4.

Draw a dot formula for the oxygen molecule, O_2.

Solution.

Since each O atom has 6 valence electrons, the only way we can account for the total (12) valence electrons and obey the octet rule is to place four of the electrons between the oxygen atoms, i.e., a double bond.

$$\dot{:}\ddot{O}{::}\ddot{O}\dot{.}$$

Example 5.

Draw a dot formula for the hydrocyanic acid molecule, HCN (covalent).

Solution.

It is necessary to know the skeleton structure, that is, which atom is bonded to which, before a dot formula can be drawn for a multi-atom compound. In this case the skeleton is H to C to N. Dot formulas for the elements are H·, ·C·, and ·N·. Thus, in the dot formula, we must account for a total of 10 valence electrons. Hydrogen has only a single valence electron and can *never* form more than a single covalent bond, i.e., there can only be two electrons between H and any atom to which it is bonded. Therefore, we can start by drawing

$$H:CN$$

We can then systematically try placing a single, double, or triple bond between C and N until the octet rule is obeyed for C and N and the total of 10 valence electrons are depicted. The correct formula must have a triple bond between C and N

$$H:C:::N:$$

Example 6.

Draw a dot formula for the sulfite ion, SO_3^{2-}.

Solution.

The bonding in the SO_3^{2-} ion is covalent. The skeleton structure is

$$\begin{array}{cc} O & O \\ S & \\ O & \end{array}$$

with each O atom bonded to S.

Then we need the total number of valence electrons. Sulfur has 6 valence electrons and each oxygen atom has 6 valence electrons for a total of $4 \times 6 = 24$. In addition, there are 2 "extra" electrons that give the -2 charge on the sulfite ion. Thus, we must place a total of 26 electrons in the dot formula for the SO_3^{2-} ion. Some possible structures you might write are

(1) :Ö:S:Ö: (2) :Ö:S:::O: (3) :Ö::S::Ö:
        :Ö: :Ö: :Ö:

(4) :Ö:S::Ö: (5) :Ö::S::Ö:
        :Ö: :Ö:

Which is the correct formula? In (1) the octet rule is obeyed and there are a total of 26 electrons. In (2) there are only 24 electrons and 10 around sulfur! What is wrong with structures (3), (4), and (5)?
The correct formula for SO_3^{2-} is (1),

$$\left[\begin{matrix} :\ddot{O}:\ddot{S}:\ddot{O}: \\ :\ddot{O}: \end{matrix} \right]^{2-}$$

Example 7.

Draw an electron dot formula for the nitrite ion, NO_2^-.

Solution.

The bonding in the NO_2^- ion is covalent, the skeleton consists of both O atoms bonded to the N atom, and there is one "extra" electron to be accounted for in addition to the five valence electrons of nitrogen and six from each oxygen. The total number of dots must be 18. The correct structure is

$$\ddot{N}::\ddot{O}: \quad \text{or} \quad \ddot{N}:\ddot{O}:$$
$$:\ddot{O}: \qquad\qquad :\ddot{O}:$$

When two or more equivalent electron dot formulas can be drawn for a compound, a condition of "resonance" is said to exist. The concept of resonance is a result of the oversimplification of using dot formulas to depict chemical bonding. Dot formulas afford a good and easy means of understanding chemical bonding, but when resonance exists it is necessary to visualize bonding as a "hybrid" — meaning a blend, or something in between. Thus, in the NO_2^- ion we must think of each N to O bond as a cross between single and double bonds. Just as the infamous Wyoming "Jackalope" is supposed to be a cross between a jack rabbit and an antelope, we must think of some chemical bonds as hybrids. In truth, the resonating chemical bond and the Jackalope probably have one thing in common: they are both fictitious!

DETERMINING SHAPES OF MOLECULES   CATEGORY
III

Most physical and many chemical properties of covalent molecules are a result of the geometry of the molecules. Water is what it is by virtue of its physical shape as much as its chemical composition. Water molecules are bent and thus are polar, having a positive and negative end. If H_2O molecules were linear and non-polar, like CO_2 molecules, the world we live in would not exist! An understanding of chemistry must include an appreciation of how molecules are shaped and how molecular geometry is determined. An effective means of predicting molecular shapes is based on dot formulas. It is often called "Valence Shell Electron Pair Repulsion," or VSEPR Theory. The idea is to consider a pair of electrons as a point or center of negative charge. Since points of negative charge will repel each other, one need only imagine how 2, 3, 4, 5, or 6 centers of negative charge will be positioned around a sphere (central atom) in order to minimize negative charge repulsion. Two, three, and four electron pairs would be positioned as follows:

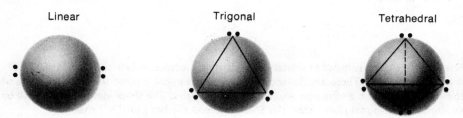

Linear   Trigonal   Tetrahedral

Five and six electron pairs will be arranged in a trigonal bipyramid and an octahedron, respectively.

To predict molecular shapes, follow two steps: (1) draw the dot formula, and count the number of electron pairs around the central atom; and (2) position the electron pairs as far apart as possible.

Example 1.

Determine the shape of the methane molecule, CH_4.

Solution.

The dot formula for CH_4 is

$$H:\overset{\displaystyle H}{\underset{\displaystyle H}{C}}:H$$

There are four pairs of electrons around the central carbon atom, so the four electron pairs will be at the four corners of a tetrahedron. All four pairs are "bond pairs" and all four electron pair repulsions are of equal intensity; therefore, the methane molecule is a perfect tetrahedron:

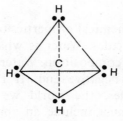

Example 2.

Determine the shape of the $SnCl_2$ molecule.

Solution.

The dot formula is

$$\overset{\displaystyle \cdot\cdot}{Sn}$$
$$:\overset{\cdot\cdot}{Cl}: \qquad :\overset{\cdot\cdot}{Cl}:$$

Many, but usually somewhat unstable, covalent molecules violate the octet rule. $SnCl_2$ is a good example. There are 6 electrons (3 electron pairs) around the Sn atom. One of the electron pairs is a lone pair and two are bond pairs. The three electron pairs will be arranged in a trigonal plane, thus the $SnCl_2$ *molecule* is bent, or "V" shaped:

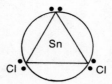

Example 3.

Determine the shape of a water molecule.

Solution.

The dot formula for H_2O is

$$\ddot{\text{:O:}}$$
$$H \quad \quad H$$

There are four pairs of electrons surrounding the oxygen atom; consequently, the four electron pairs will occupy the corners of a tetrahedron:

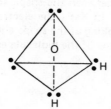

Two of the electron pairs are bond pairs and two are lone pairs. Since all four corners of a tetrahedron are equivalent, it is completely arbitrary which two corners are occupied by H atoms. It is important to note that while the four electron pairs, lone pairs, and bond pairs describe a tetrahedron, the shape of the H_2O *molecule* would not be tetrahedral. It would simply be described as "bent" or angular. A point of note here is that lone pairs repel slightly more than bond pairs. Thus, in the H_2O molecule the lone pairs are spread apart and the bond pairs are squeezed together. In a perfect

tetrahedron, like the CH_4 molecule, the H—C—H bond angle is 109°: $C \overset{H}{\underset{H}{\diagdown}}$ 109°,

but in H_2O, the H—O—H bond angle is only 103°. This fact contributes to the degree to which the water molecule is bent, and thus to the total polarity of the molecule.

Example 4.

Determine the shape of xenon tetrafluoride, XeF_4.

Solution.

In the electron dot formula for XeF_4 there is necessarily a gross violation of the octet rule. Xenon has 8 valence electrons and each fluorine atom must share one of those, so we end up with 12 electrons (6 e⁻ pairs) around the xenon atom:

Six electron pairs will be arranged in an octahedron around the xenon atom. All corners of an octehedron are equivalent, but lone pairs repel to a greater degree than bond pairs and the two lone pairs will be positioned as far apart as possible. The structure is

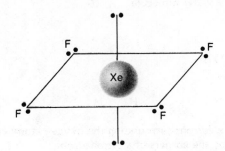

Once again, note the difference between the shape adopted by *all* electron pairs and the actual shape of the *molecule.* In XeF_4 the electron pairs are directed toward the corners of an octahedron, but looking at the shape adopted by the xenon and fluorine atoms, we would describe the shape of the molecule as flat and square (square planar). Would XeF_4 be a polar molecule?

PROBLEMS   *Category I*

1. Write electron configurations for the following elements:
   (a) F   (b) Si   (c) K   (d) Mn   (e) Ir   (f) Te

2. Write electron configurations for the following ions:
   (a) Cl^- (b) Ca^{2+} (c) Fe^{2+} (d) Cr^{3+} (e) N^{3-} (f) Hg^{2+}

3. Which atoms, or ions, have the following electron configurations?
   (a) (Kr core) $5s^1$
   (b) (Ar core) $3d^{10} 4s^2 4p^3$
   (c) (Ar core) $3d^7$
   (d) $1s^2 2s^2 2p^6 3s^2 3p^6$
   (e) (Ar core) $3d^{10} 4s^2$
   (f) (Xe core) $4f^{14} 5d^3 6s^2$

4. Identify the first element appearing in the periodic table which has
   (a) four 2p electrons.
   (b) two 4p electrons.
   (c) two electrons in a d subshell.
   (d) three electrons in the 6p subshell.
   (e) one electron in the 4d subshell.

5. Write the electron configuration for the (as yet hypothetical) element with atomic number 118.

Category II

6. Draw electron dot formulas for
   (a) HCl   (b) CO   (c) NO^+ (d) CN^- (e) N_2

7. Predict the formulas for the ionic compounds formed between
   (a) Mg and O
   (b) K and O
   (c) Al and O
   (d) Ca and N
   (e) Cs and Se

8. Using the skeleton structures provided, draw dot formulas for each of the following:

(a) C_2H_4

H H
C C
H H

(b) NO_3^-

O   O
   N
   O

(c) C_6H_6

          H
          C
H   C   C   H
H   C   C   H
          C
          H

(d) HSCN

H S C N

(e) $PO_3{}^{3-}$

O   O
   P
   O

9. Of the structures in Problem 8, three exhibit resonance. Which three?

10. Based on dot formulas, which of the following do not obey the octet rule?

(a) OF_2 (b) BF_3 (c) NO   (d) CO_2 (e) PCl_5 (f) C_2H_2

Category III

11. Determine the shapes of

(a) OF_2 (b) BF_3 (c) PCl_5

12. Predict an approximate H—N—H bond angle in the ammonia molecule, NH_3.

13. $BeCl_2$ and $TeCl_2$ are both covalent molecules, yet $BeCl_2$ is linear while $TeCl_2$ is "V" shaped. Explain.

14. SF_6 is a perfect octahedron but XeF_6 is a distorted octahedron. Explain.

15. Determine the shapes of

(a) $GeCl_2$ (b) PCl_3 (c) IF_4^- (d) $SeCl_4$ (e) IF_5

General Problems

16. What is the most likely formula of a covalent molecule formed from
(a) an element in group VA with an element in group VIIA?
(b) an element in group IVA with an element in group VIA?

17. Based upon electron configurations and the octet rule, explain the following:
(a) VF_5 exists but VF_6 does not.
(b) $TiCl_4$ is much more stable than $TiCl_3$.
(c) carbon forms two types of carbide ions, C^{4-} and $C_2{}^{2-}$.

18. Write electron configurations for
(a) Te^{2-} (b) Lu   (c) Pt^{6+} (d) U^{6+}

19. Draw electron dot formulas for
(a) Al_2F_6 (b) PbI_3^- (c) SO_3Cl^- (d) NSCl

20. Which of the following would you expect to have shapes that would make them polar molecules?
(a) SiF_4 (b) XeF_4 (c) SF_4 (d) PF_5 (e) NF_3

21. Name the element that fits the description:
(a) the first atom in the periodic table with five electrons in a p subshell.
(b) the first noble gas with electrons in a d subshell.
(c) the first atom with an electron in the 5s subshell.
(d) the element whose +2 ion has an Ar configuration.

22. Give the formula of a polyatomic molecule that would have the same shape as
(a) H_2O (b) PCl_3 (c) IF_3 (d) GeH_4

6 LIQUIDS, SOLIDS, AND PHASE CHANGES

While distances between molecules are very large in gases, intermolecular distances must be relatively small in liquids and solids. Gases are characterized by large volumes and rapid, random molecular motion. Solids are characterized by close packing of atoms or molecules in orderly arrangements called crystal structures, in which there are only slight vibrational, molecular motions. Liquids may be visualized as either condensed gases or melted solids, with intermediate molecular motions.

The problems in this brief chapter are divided into three categories.

Problem Categories

    I   Vapor pressures of liquids (the Clausius-Clapeyron equation).
   II   Some calculations involving solid crystals.
  III   Phase changes.

VAPOR PRESSURES OF LIQUIDS (THE CLAUSIUS-CLAPEYRON EQUATION

CATEGORY I

Atoms, molecules, or ions are held together by various attractive forces in the liquid phase, yet there must be considerable movement of particles in a liquid. What happens if a drop of ink is gently placed on the surface of a glass of liquid water? The liquids mix. The ink spreads (diffuses) throughout the water rather rapidly. The molecules making up the ink, and the water molecules must both be in motion. Molecules in motion have kinetic energy (K.E. = $\frac{1}{2} mv^2$). In a body of a liquid some of the molecules near the surface of the liquid have sufficient kinetic energy to overcome the forces of attraction holding them together and escape the surface of the liquid. This is the familiar process of evaporation. In an open container, the molecules entering the vapor phase are free to move out into the atmosphere. What happens in a closed container? Is there evaporation in a stoppered flask? Yes, there is, but in a closed container the vapor phase soon becomes saturated with molecules and the rate of condensation equals the rate of evaporation. When this occurs there is said to be a state of equilibrium between the vapor and the liquid. Like any gas, the vapor in equilibrium with its liquid will exert a pressure and that pressure is called the *vapor pressure* of the liquid.

Vapor pressure is a function of (1) the energy necessary to overcome attractive forces in the liquid and (2) the average kinetic energy of the molecules. The energy needed to overcome attractive forces in the liquid is measured by the heat of vaporization. The *molar heat of vaporization*, ΔH_v, is the amount of heat required to vaporize one mole of liquid against a constant pressure. The average kinetic energy of a collection of molecules is directly proportional to the temperature, in °K. So, vapor pressure is a function of the parameters ΔH_v and T. This relationship is concisely stated in an equation known as the Clausius-Clapeyron equation:

$$\log \frac{P_2}{P_1} = \frac{\Delta H_v \text{ (cal/mole)}}{4.58 \text{ cal/mole-}^\circ K} \left(\frac{T_2 - T_1}{T_2 T_1} \right)$$

P_1 is the vapor pressure of a liquid at temperature T_1, and P_2 is the vapor pressure at a temperature T_2. ΔH_v is the molar heat of vaporization and the number 4.58 is a constant. Let's see how this important equation is applied to liquids.

The *boiling point* of a liquid is defined as that temperature at which its vapor pressure becomes equal to the pressure above its surface (usually atmospheric pressure).

The volatility of a liquid, that is, how readily it vaporizes, can be described by (1) its standard boiling point (external pressure = 1 atm), (2) its molar heat of vaporization, or (3) its vapor pressure at a particular temperature.

Example 1.

The molar heat of vaporization of water is 9700 cal/mole. The vapor pressure of water is 17.5 mm Hg at 20°C. Find the vapor pressure of water at 80°C.

Solution.

This is a "plug into the equation" type of problem. If we call P_1 17.5 mm Hg and T_1 20°C (293°K), then T_2 is 80°C (353°K), and we can substitute into the Clausius-Clapeyron equation.

$$\log \frac{P_2}{P_1} = \frac{\Delta H_v \text{ (cal/mole)}}{4.58 \text{ cal/mole-}^\circ K} \left(\frac{T_2 - T_1}{T_2 T_1} \right)$$

$$\log \frac{P_2}{17.5 \text{ mm Hg}} = \frac{9700 \text{ cal/mole}}{4.58 \text{ cal/mole-}^\circ K} \left(\frac{353^\circ K - 293^\circ K}{353^\circ K \times 293^\circ K} \right)$$

$$\log \frac{P_2}{17.5 \text{ mm Hg}} = 1.23$$

$$\log P_2 - \log 17.5 = 1.23$$

$$\log P_2 - 1.24 = 1.23$$

$$\log P_2 = 2.47$$

$$P_2 = 295 \text{ mm Hg}$$

For manipulation of logarithms, see Appendix I. Notice that P can be in mm Hg, atm, etc.

Example 2.

The molar heat of vaporization of diethyl ether is 6200 cal/mole. The vapor pressure of ether is 442 mm Hg at 20°C. Find the standard (1 atm pressure) boiling point of ether.

Solution.

We can call P_1 442 mm Hg, P_2 760 mm Hg, T_1 20°C (293°K), and solve for the unknown T_2 in the Clausius-Clapeyron equation.

$$\log \frac{760 \text{ mm Hg}}{442 \text{ mm Hg}} = \frac{6200 \text{ cal/mole}}{4.58 \text{ cal/mole-}°\text{K}} \left(\frac{T_2 - 293°\text{K}}{T_2 \times 293°\text{K}} \right)$$

$$\log 1.72 = 1354 \left(\frac{T_2 - 293°\text{K}}{T_2 \times 293°\text{K}} \right)$$

The log of 1.72 is 0.236; therefore,

$$\frac{0.236}{1354} = \left(\frac{T_2 - 293°\text{K}}{T_2 \times 293°\text{K}} \right),$$

and $0.051 \, T_2 = T_2 - 293°\text{K}$.

$$-0.949 \, T_2 = -293°\text{K}$$

$$T_2 = 309°\text{K or } 36°\text{C}$$

Ether is a very volatile liquid with a high vapor pressure and low standard boiling point compared to water.

Example 3.

At the summit of Mt. Everest (29,000 ft above sea level) the atmospheric pressure was measured at 220 mm Hg. At what temperature would water boil under that pressure?

Solution.

The standard boiling point of water (760 mm Hg) is 100°C or 373°K. We can substitute into the Clausius-Clapeyron equation using T_1 = 373°K and P_1 = 760 mm Hg. P_2 will be the atmospheric pressure atop Mt. Everest, 220 mm Hg. ΔH_v (H_2O) = 9700 cal/mole.

$$\log \frac{220 \text{ mm Hg}}{760 \text{ mm Hg}} = \frac{9700 \text{ cal/mole}}{4.58 \text{ cal/mole-}°\text{K}} \left(\frac{T_2 - 373°\text{K}}{T_2 \times 373°\text{K}} \right)$$

Solving as in Example 2, we find

$$T_2 = 341°\text{K or } 68°\text{C}$$

If you ever climb Mt. Everest, a bath in boiling water would be welcome and quite comfortable when you reach the summit. At what temperature would water boil on top of Pike's Peak in Colorado (elevation: 14,110 ft above sea level, atmospheric pressure ~ 450 mm Hg)?

CATEGORY   **SOME CALCULATIONS INVOLVING SOLID CRYSTALS**

II

A solid is defined as a solid because it is crystalline. The atoms, ions, or molecules in a solid will arrange themselves in definite patterns called crystal structures. Most solids crystallize in structures consistent with a "close packing" of atoms. If we think of atoms, molecules, or ions as spheres and imagine how spheres could be stacked in three dimensions, we can begin to visualize crystal structures. If we then try to imagine stacking spheres of different sizes and spheres that are positively and negatively charged, the problem soon gets quite complicated. Solid crystals exist in many different shapes and structures. Depending upon the nature of the "spheres," solid crystals may be classified as molecular (e.g., ice), covalent (e.g., diamond), metallic (any metal), or ionic (e.g., NaCl). There are seven different simple crystal systems labeled cubic, tetragonal, hexagonal, rhombohedral, orthorhombic, monoclinic, and triclinic. There are variations within some of those categories. Volumes have been written about crystal structures, and it is impossible here to go into much detail. Let's concentrate on the three variations of the cubic system. The three unit cells of the cubic system are shown in Figure 6–1. The *unit cell* is the smallest unit, which if repeated in three dimensions, will generate the entire crystal structure.

With a knowledge of crystal structure (from, e.g., X-ray diffraction studies), density, atomic, and ionic sizes, etc., we can perform some interesting calculations on crystals.

Example 1.

Magnesium oxide crystallizes in a simple cubic structure. Each unit cell contains four

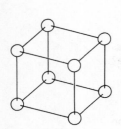

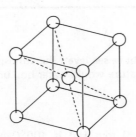

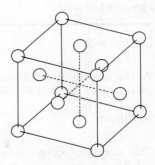

Simple cubic

Body-centered cubic

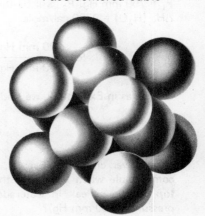
Face-centered cubic

Mg^{2+} ions and four O^{2-} ions. The density of crystalline MgO is 3.62 g/cm³ at 25°C. Find the distance between Mg^{2+} and O^{2-} ions (i.e., the length of one side of the cube).

Solution.

First, let's find the weight of 4 Mg^{2+} ions and 4 O^{2-} ions.

$$X \text{ g} = 4 \text{ } Mg^{2+}O^{2-} \times \frac{1 \text{ mole } Mg^{2+}O^{2-}}{6.02 \times 10^{23} \text{ } Mg^{2+}O^{2-}} \times \frac{40.3 \text{ g (MgO)}}{1 \text{ mole } Mg^{2+}O^{2-}} = 2.68 \times 10^{-22} \text{g}$$

If the density is 3.62 g/cm³, then the volume occupied by 2.68×10^{-22} g is

$$X \text{ cm}^3 = 2.68 \times 10^{-22} \text{ g} \times \frac{1 \text{ cm}^3}{3.62 \text{ g}} = 7.40 \times 10^{-23} \text{ cm}^3$$

The length of one side of the tiny cube is $\sqrt[3]{7.40 \times 10^{-23}} = 4.20 \times 10^{-8}$ cm. One angstrom is 10^{-8} cm, so the length of the simple cubic MgO unit cell is 4.20 angstroms.

Example 2.

Metallic gold crystallizes in a face-centered cubic structure which is 4.078 angstroms on an edge. The density of gold is 19.30 g/cm³ at 25°C. Find (a) the atomic radius of gold, and (b) Avogadro's number.

Solution.

(a) In a face-centered cube, the corner atoms are touching the face center atom, but not each other. Looking at one face we would see an arrangement similar to that shown in Figure 6–2.

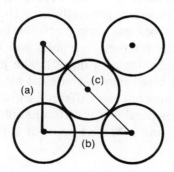

If distance a (or b) is 4.078 Å, we can find distance c from the Pythagorean theorem:

$$a^2 + b^2 = c^2$$

$$(4.078)^2 + (4.078)^2 = c^2$$

$$c^2 = 33.26 \text{ Å}^2 \text{; so } c = 5.767 \text{ Å}$$

Examining the figure we see that distance c would be the sum of the diameters of two gold atoms. The radius of a single atom would be c ÷ 4, or 5.767 Å ÷ 4 = 1.442 Å.

(b) How many atoms are there in a cube of gold 4.078 Å on a side? If we look at a picture of the face-centered cubic unit cell and count, we see 14. But, those 14 are being shared by adjacent unit cells. Each atom on a face is being shared by two unit cells and each corner atom is actually the corner of 8 different cubes. There are 6 "face" atoms, only half of which belong to a particular unit cell (½ × 6 = 3) and there are 8 corner atoms, only $\frac{1}{8}$ of which belong to a given unit cell ($\frac{1}{8}$ × 8 = 1). So, in the face-centered cube we can actually assign only 4 atoms to each unit cell.

Now, we can find Avogadro's number. In a cube 4.078 Å (4.078 × 10^{-8} cm) on a side there are 4 atoms:

$$(4.078 \times 10^{-8} \text{ cm})^3 = 6.782 \times 10^{-23} \text{ cm}^3 = 4 \text{ atoms}$$

The density of gold is 19.30 g/cm^3. How many atoms are there in 1 mole (197.0 g) of gold?

$$X \text{ atoms} = 197.0 \text{ g Au} \times \frac{1 \text{ cm}^3}{19.30 \text{ g Au}} \times \frac{4 \text{ atoms}}{6.782 \times 10^{-23} \text{ cm}^3} = 6.020 \times 10^{23} \text{ atoms}$$
$$\text{(per mole)}$$

Calculations of Avogadro's number from crystal structure data are usually in good agreement with the accepted value.

CATEGORY III   PHASE CHANGES

All substances can exist in the solid, the liquid, or the gas phase, or any two phases simultaneously, or even all three phases simultaneously. How a substance exists depends only on the temperature and pressure. We usually think of something like metallic iron only as a solid, but at a high enough temperature and/or a low enough pressure, iron can be liquid or gaseous. Conversely, we think of oxygen as a gas, but at low temperatures and high pressures oxygen can be a liquid or a solid. The temperature-pressure dependence of phase is best described by a *phase diagram*. Phase diagrams generally have the form shown in Figure 6–3. The line separating liquid from gas represents the boiling point of a liquid as a function of pressure and the line between solid and liquid describes the melting point of a solid (or freezing point of a liquid). The junction of the lines is called the triple point, where all three phases can exist simultaneously. The form of a phase diagram will vary somewhat for different substances. The temperature and pressure axes may vary considerably for different substances. For example, water has a melting point of 0°C and a boiling point of 100°C at 760 mm Hg while the

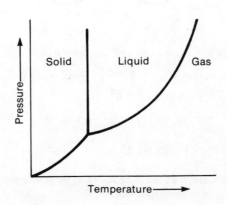

triple point for water occurs at 0.01°C and 4.56 mm Hg. By comparison, the triple point for carbon dioxide occurs at a temperature of −57°C and a pressure of 3876 mm Hg (5.1 atm).

The melting of a solid or boiling of a liquid are endothermic processes. How readily a solid melts or a liquid boils at constant pressure is reflected by its *heat of fusion* or *heat of vaporization.* The heat required to melt ("fuse") one mole of water at 0°C and 1 atm is ΔH_f = 1.44 kcal/mole. The heat required to boil (vaporize) one mole of water at 100°C and 1 atm is ΔH_v = 9.72 kcal/mole. Heats of fusion and vaporization are usually listed in kcal per mole or calories per gram. Some typical values of ΔH_f and ΔH_v are given in Table 6–1 in cal/g, at 1 atm pressure and the temperatures indicated.

Table 6–1 Typical Values of ΔH_f and ΔH_v for Selected Substances

Substance	ΔH_f (cal/g)	ΔH_v (cal/g)
water	80 (0°C)	540 (100°C)
NaCl	116 (808°C)	698 (1465°C)
HCl	13.1 (−114°C)	106 (−85°C)
benzene	32.7 (5°C)	94 (80°C)

Let's look at a few problems involving ΔH for phase changes.

Example 1.

Calculate the heat energy required to vaporize 10.0 g of benzene at its boiling point, 80° (1 atm).

Solution.

Since we know ΔH_v in cal/g, we can solve:

$$X \text{ cal} = 10.0 \text{ g benzene} \times \frac{94 \text{ cal}}{1 \text{ g benzene}} = 940 \text{ cal}$$

Example 2.

How many kcal are required to completely convert 100 g of ice at 0°C into steam at 100°C?

Solution.

First, we can find ΔH for melting 100 g of ice at 0°C.

$$X \text{ cal} = 100 \text{ g ice} \times \frac{80 \text{ cal}}{1 \text{ g ice}} = 8000 \text{ cal} = 8 \text{ kcal}$$

How many kcal will be needed to heat 100 g of liquid water from 0°C to 100°C? The "specific heat" of liquid water is 1.00 cal/g-°C; that is, it takes 1.00 cal to raise the temperature of 1.0 g of liquid water by 1.0°C. Therefore, we can find

$$X \text{ cal} = 100°C \times \frac{100 \text{ cal}}{1.0°C} = 10,000 \text{ cal} = 10 \text{ kcal}$$

Note: For 100 g H_2O, a 1.0°C temperature rise would be equivalent to 100 cal, hence, the conversion factor.

Finally, we must find ΔH for converting 100 g of liquid water to gaseous water (steam) at 100°C. Since we know the heat of vaporization, we can solve:

$$X \text{ cal} = 100 \text{ g water} \times \frac{540 \text{ cal}}{1 \text{ g water}} = 54{,}000 \text{ cal} = 54 \text{ kcal}$$

The total number of kcal needed to go from ice to steam = 8 + 10 + 54 = 72 kcal.

Example 3.

If 10.0 g of ice at 0°C is placed in 100 g of water at 25°C, to what temperature could the original 100 g of water be lowered by the melting of the ice?

Solution.

From the ΔH_f of ice we find that melting 10.0 g of ice would require

$$X \text{ cal} = 10.0 \text{ g ice} \times \frac{80 \text{ cal}}{1 \text{ g ice}} = 800 \text{ cal}$$

If that much heat is transferred from the water to the ice, what will happen to the temperature of the water? Removing 800 cal from 100 g of water would lower the temperature by 8°C, hence the temperature would go from 25°C to 17°C. Would 17°C be the final temperature of the 100 g of water plus the 10.0 g of melted ice?

PROBLEMS   *Category I*

1. The molar heat of vaporization of carbon tetrachloride, CCl_4, is 7.17 kcal/mole. The vapor pressure of CCl_4 is 760 mm Hg at 76.7°C. Find the vapor pressure of CCl_4 at room temperature (25°C).

2. Propene, C_3H_6, has a vapor pressure of 198 mm Hg at 200°K and 10,040 mm Hg at 300°K. Calculate the molar heat of vaporization of propene.

3. The molar heat of vaporization of ethyl alcohol is 9.22 kcal/mole. The standard boiling point of ethyl alcohol is 78.5°C. Find the boiling point of ethyl alcohol on top of Pike's Peak (atmospheric pressure 450 mm Hg).

4. The atmospheric pressure on the planet Mars is 5.7 mm Hg. At what temperature would water boil on Mars? $\Delta H_v = 9700$ cal/mole.

5. Would you expect to find liquid water on the surface of Venus, where the atmospheric pressure is 50 atm and the temperature is 700°K? Prove your answer.

Category II

6. How many atoms can actually be assigned to a unit cell in the (a) body-centered cube? (b) simple cube?

7. Nickel metal crystallizes in a face-centered cubic structure. The atomic radius of Ni is 1.25 Å. What is the length of one side of a unit cell in a Ni crystal?

8. Strontium oxide, SrO, crystallizes in a simple cubic structure with 4 Sr^{2+} and 4 O^{2-} making up each simple cube. The length of the unit cell is 5.15 Å. Find the density of SrO.

9. Cesium metal crystallizes in a body-centered cubic structure. Each corner atom is in contact with the "center" atom, but not with each other. The length of one side of the unit cell is 6.08 Å. Find the radius of the Cs atom.

10. In a diamond crystal each carbon atom is surrounded tetrahedrally by 4 other C atoms. Draw a unit cell. How many C atoms should be assigned to the unit cell?

Category III

11. Sublimation is conversion of a substance from the solid phase directly to the gas phase. From the information in Table 6–1 find the heat change for the sublimation of 1 mole of water (the heat of sublimation).

12. A substance has a standard (1 atm) melting point of 65°C and standard boiling point of 140°C. The triple point occurs at 55°C and 10 mm Hg pressure. Sketch a phase diagram for the substance. Describe what happens when the solid at 40°C and 5 mm Hg pressure is heated at constant pressure. What will happen if the substance is in the liquid phase at 60°C and 50 mm Hg and the pressure is increased to 760 mm Hg while the temperature is held constant?

13. Find ΔH for (a) melting 15.6 g of benzene at 5°C; (b) boiling 15.6 g of benzene at 80°C.

14. If 15.6 g of solid benzene at 5°C is immersed in 100 g of water at 20°C, what will be the final water temperature?

15. If you had 1.0 liter of water at room temperature (25°C), how much heat would have to be removed to convert the water to ice at 0°C?

General Problems

16. It is found that 59 calories are required to vaporize 1.0 g of chloroform, $CHCl_3$, at its standard boiling point, 62°C. Find the vapor pressure of chloroform at 20°C.

17. It has been suggested that beneath the atmosphere of the planet Jupiter there is a sea of liquid ammonia. The standard boiling point of NH_3 is –33°C and the molar heat of vaporization is 5560 cal/mole. If the surface temperature of Jupiter were 600°K, what would the minimum atmospheric pressure have to be in order for such a sea to exist?

18. Sodium metal crystallizes in a body-centered cubic structure. The atomic radius of Na is 1.86 Å and the density of metallic sodium is 0.971 g/cm³ at 0°C. Calculate Avogadro's number. (What assumptions must you make?)

19. Rubidium iodide, RbI, crystallizes in a body-centered cubic structure with each Rb^+ ion surrounded by 8 I^- ions, and vice-versa. The ionic radius of Rb^+ is 1.49 Å and the ionic radius of I^- is 2.17 Å. Find the length of one side of the RbI unit cell.

20. A pan of water placed on a stove contains 400 g of water at 20°C. How many calories will be required to completely boil away the water?

7 CONCENTRATION AND PROPERTIES OF SOLUTIONS

The term "concentration" refers to the amount of one substance present for a given amount of "something else." Usually concentrations are used to describe the amount of a solute present in some quantity of solvent or solution, but not always. The "amount of substance" may be in grams, atoms, molecules, moles, pounds, ounces, milliliters, liters, quarts, equivalents, etc., and the amount of "something else" may be liters of solution, grams of solvent, total moles of all species present, volume of a container, or total mass of solution, to name a few! There are literally hundreds of ways of describing concentration. Let us concentrate (no pun intended) on a few of the terms most valuable to the chemist.

A *solution* is any homogeneous mixture of two or more substances. In a solution the *solvent* is the component in excess and may contain one or more *solutes*. By far the most widely used concentration term in chemistry is *molarity*. Molarity, given the symbol M, is defined as *moles of solute per liter of solution*. Stated as a mathematical expression:

$$M = \frac{\text{moles solute}}{\text{liters solution}}$$

If 1.0 liter of a solution contained 0.50 mole of a strong electrolyte like $AgNO_3$, the solution is 0.50 molar in $AgNO_3$. Since $AgNO_3$ is completely ionized in solution, it could also be described as 0.50 molar (0.50 M) in Ag^+ and 0.50 M in NO_3^-.

Problem Categories

There are three variables in the definition of molarity. Due to the importance of this particular concentration term, the first three categories in this chapter are devoted to molarity problems in which we solve respectively for molarity, number of moles of solute and number of liters of solution.

I Given the amount of solute and volume of solution, what is the molarity of the solution?

II Given the molarity and a volume of solution, how many moles or grams of solute does that volume of solution contain?

III Given the molarity of a solution, what volume of the solution will contain a known number of moles or grams of solute?

IV Other concentration terms.

V Some effects of concentration on solution properties.

MOLARITY FROM AMOUNT OF SOLUTE AND VOLUME OF SOLUTION CATEGORY

Example 1.

If 0.10 mole of NaCl is dissolved in enough water to give 500 ml of solution, what is the molarity of the solution?

Solution.

The problem can be solved by either of two methods. One we will call the "definition" method and the other is the factor-label method. The definition method consists simply of substitution of the known quantities into a definition and solving for the unknown quantity. Thus, since we know the number of moles of solute (0.10) and the number of liters of solution (0.500), we can easily find the molarity:

$$M = \frac{\text{moles solute}}{\text{liters solution}} = \frac{0.10 \text{ mole NaCl}}{0.500 \text{ liter solution}}$$

$$M = 0.20 \text{ mole NaCl per liter of solution}$$

By the factor-label method, we ask the question, "How many moles of NaCl are equivalent to one liter of solution?", or

$$X \text{ moles NaCl} = 1.0 \text{ liter solution}$$

For this particular NaCl solution, we know that 0.10 mole of NaCl is equivalent to 500 ml (0.500 ℓ) of solution, and that gives us the necessary unity factor:

$$X \text{ moles NaCl} = 1.0 \text{ liter solution} \times \frac{0.10 \text{ mole NaCl}}{0.500 \text{ liter solution}} =$$

$$0.20 \text{ mole NaCl (per liter of solution)}$$

Either method can be used. Throughout this chapter we will work with the "definition" method.

Keep in mind that molarity refers to moles of solute per liter of *solution*. To find molarity we must know the exact volume of the solution, not just the volume of the solvent. If this problem had read "0.10 mole of NaCl dissolved in 500 ml of water," we could not have calculated the molarity exactly. The volume of water may be exactly 500 ml, but what would the volume of solution be after the NaCl is added to the water? It could be 503 ml; it could be 498 ml. We would not know for sure, and to find the molarity of the solution the final total volume must be known.

Example 2.

If 7.30 grams of gaseous hydrogen chloride is dissolved in water to give 250 ml of solution, calculate the molarity of the HCl solution.

Solution.

The number of moles of a substance is the weight divided by the formula weight. In this case,

$$\text{moles HCl} = \frac{\text{wt HCl}}{\text{MW HCl}} = \frac{7.30 \text{ grams}}{36.5 \text{ grams/mole}} = 0.20 \text{ mole}$$

and from the definition of molarity,

$$M = \frac{0.20 \text{ mole HCl}}{0.250 \text{ liter solution}} = 0.80 \text{ mole per liter}$$

Example 3.

When 50.0 ml of water and 150.0 ml of ethyl alcohol, C_2H_5OH, are mixed at 25°C, the volume of the final solution is 195.0 ml. Calculate the molarity of the water (density = 1.00 g/ml).

Solution.

First, decide which is the solute and which the solvent. Since the alcohol is in excess we must view this as a solution of H_2O (the solute) dissolved in alcohol (the solvent). Note that the volumes are not additive. This is often the case when two liquids are mixed.

If we take the density of water to be 1.00 gram per ml, then 50.0 ml of H_2O = 50.0 grams of H_2O, and the number of moles will be

$$\text{moles} = \frac{\text{wt } H_2O}{\text{MW } H_2O} = \frac{50.0 \text{ grams}}{18.0 \text{ grams per mole}} = 2.78 \text{ moles}$$

Then, from the definition of molarity,

$$M = \frac{2.78 \text{ moles } H_2O}{0.195 \text{ liter solution}} = 14.3 \text{ moles per liter}$$

CATEGORY II   **AMOUNT OF SOLUTE FROM MOLARITY AND VOLUME OF SOLUTION**

Example 1.

How many moles of sodium hydroxide, NaOH, are there in 25.0 ml of 0.15 M NaOH solution?

Solution.

From the definition:

$$M = \frac{\text{moles solute}}{\text{liters solution}} \text{ ; moles solute} = M \times \text{liters solution}$$

$$\text{moles NaOH} = 0.15 \text{ M} \times 0.025 \text{ } \ell = 3.8 \times 10^{-3}$$

Example 2.

If the acid in a lead storage cell is 5.0 M H_2SO_4, how many grams of H_2SO_4 are contained in 10.0 ml of this "battery acid"?

Solution:

First, determine the number of moles of H_2SO_4 from

$$\text{moles } H_2SO_4 = 5.0 \text{ M} \times 0.010 \text{ liter} = 0.050 \text{ mole}$$

Then knowing that

$$\text{moles } H_2SO_4 = \frac{\text{Wt } H_2SO_4}{\text{MW } H_2SO_4}$$

we can solve for the weight of H_2SO_4:

$$\text{wt } H_2SO_4 = 0.050 \text{ mole} \times 98.0 \text{ grams/mole} = 4.9 \text{ grams}$$

Example 3.

After 10.0 ml of 5.0 M H_2SO_4 is diluted to a volume of 200 ml by addition of water, how many moles and grams of H_2SO_4 are there in 20.0 ml of the dilute solution?

Solution.

Using the definition method, we must first calculate the molarity of the dilute solution. We know the number of moles of H_2SO_4 available in 10.0 ml of 5.0 M H_2SO_4 is 0.05 (see Example 2), and that the final volume of the diluted solution is 200 ml (0.200 liter), so the molarity of the diluted solution is

$$M = \frac{0.050 \text{ mole}}{0.200 \text{ liter}} = 0.25 \text{ mole per liter}$$

Then, the number of moles of H_2SO_4 in 20.0 ml of the dilute solution is found by

$$\text{moles } H_2SO_4 = M \times \text{liters} = 0.25 \text{ M} \times .020 \text{ liter} = 0.005 = 5.0 \times 10^{-3} \text{ mole}$$

The number of grams of H_2SO_4 corresponding to 5.0×10^{-3} mole is

$$\text{wt } H_2SO_4 = \text{moles} \times \text{MW} = 5.0 \times 10^{-3} \times 98.0 = 0.49 \text{ grams}$$

VOLUME OF SOLUTION FROM MOLARITY AND AMOUNT OF SOLUTE   CATEGORY III

Example 1.

What volume of 0.25 M $Pb(NO_3)_2$ solution would be required to give 0.01 mole of Pb^{2+} ion?

Solution.

Since $Pb(NO_3)_2$ contains one mole of Pb^{2+} for one mole of $Pb(NO_3)_2$, we know that a 0.25 M $Pb(NO_3)_2$ is 0.25 M in Pb^{2+}
Given the molarity and the required number of moles, it is a simple matter to find the desired volume from the definition method:

$$\text{liters solution} = \frac{\text{moles solute}}{M} = \frac{0.01 \text{ mole } Pb^{2+}}{0.25 \text{ M}} = 0.040 \text{ liter, or 40 ml}$$

Example 2.

A solution used for intravenous feeding is 5% (0.30 M) glucose in water. How many mls of this solution would be required to deliver 8.0 grams of glucose, $C_6H_{12}O_6$?

Solution.

Using definitions, first calculate the number of moles of glucose in 8.0 grams.

$$\text{moles} = \frac{\text{wt}}{\text{MW}} = \frac{8.0 \text{ grams}}{180 \text{ grams/mole}} = 0.044 \text{ mole}$$

Then, from the molarity, obtain the number of liters needed to give 0.044 mole:

$$\text{liters solution} = \frac{0.044 \text{ mole}}{0.30 \text{ mole/liter}} = 0.15 \text{ } \ell \text{ or } 150 \text{ ml}$$

Example 3.

It is estimated that there are 10,000,000,000 tons of gold present in sea water. The "concentration" is often given as "200 parts by weight to 100,000,000 parts of water." Convert this concentration to molarity and calculate the volume of sea water needed to yield 1.0 gram of gold.

Solution.

If there are 200 grams of gold per 100,000,000 grams of water, the number of moles of Au would be:

$$\text{moles} = \frac{\text{wt Au}}{\text{formula wt}} = \frac{200 \text{ grams}}{197.0 \text{ grams/mole}} = 1.02 \text{ moles Au}$$

The density of sea water varies with temperature and dissolved mineral content. If we use 1.0 gram/ml for the density of sea water, then 1.0×10^8 grams = 1.0×10^8 mls, or 1.0×10^5 liters. The molarity of the "solution" is

$$M = \frac{1.02 \text{ moles Au}}{1.0 \times 10^5 \text{ liters}} \simeq 1.0 \times 10^{-5}$$

(We needn't worry about answers to three significant figures since our density of sea water is an estimate!)
If we need 1.0 gram of gold, that would be 1.0/197 = .005 mole of Au. And, from the definition of molarity, the number of liters of 1.0×10^{-5} M solution required to yield .005 mole of gold would be

$$\text{liters solution} = \frac{.005 \text{ mole Au}}{1.0 \times 10^{-5} \text{ mole/liter}} = 500 \text{ liters}$$

Five hundred liters is not a forbidding quantity, and as technology is developed the oceans will be "mined" for gold as well as many other elements.

Example 4.

What volume of 0.15 M HCl can be prepared from 20.0 ml of 18 M HCl?

Solution.

The number of moles of HCl available is found from

$$\text{moles HCl} = M \times \text{liters (concentrated solution)}$$

$$= 18\ M \times 0.020\ \ell = 0.36\ \text{mole}$$

Then, knowing the number of moles of HCl and the molarity of the dilute solution, we can find the volume of dilute solution.

$$\text{liters solution} = \frac{0.36\ \text{mole}}{0.15\ \text{mole/liter}} = 2.4\ \text{liters}$$

OTHER CONCENTRATION TERMS   CATEGORY
IV

While molarity is the concentration unit usually employed in general chemistry, other concentration terms are often useful. The most important of these are molality, normality, and mole fraction.

Molality

Molality, m, is defined as

$$m = \frac{\text{moles solute}}{\text{kgs solvent}}$$

Molality is often used to describe the concentration of solutions when the temperature is likely to be variable. For example, molality is the concentration unit used in studies of colligative properties of solutions (lowering of freezing points, etc.). Unlike molarity, molality has the advantage of remaining constant with changing temperature. As the temperature changes, the volume of a solution, and hence the molarity, may change, but the number of kilograms of solvent remains constant.

Example 1.

Ten grams of the "antifreeze" ethylene glycol, $C_2H_6O_2$, is dissolved in 450 grams of water. Calculate the molality of the solution.

Solution.

$$\text{moles } C_2H_6O_2 = \frac{\text{wt}}{\text{MW}} = \frac{10.0\ \text{g}}{62.0\ \text{g/mole}} = 0.16\ \text{mole}$$

$$\text{molality, } m = \frac{\text{moles } C_2H_6O_2}{\text{kgs water}} = \frac{0.16\ \text{mole}}{0.450\ \text{kg}} = 0.36\ \text{mole per kg}$$

Normality

Normality, N, is defined as

$$N = \frac{\text{equivalents solute}}{\text{liters solution}}$$

Normality is viewed by some chemists as an anachronistic concentration term, but it is useful, especially for acid-base and oxidation-reduction reactions in solution. "Equivalents" of solute may be defined as

$$\text{equivalents} = \frac{\text{wt solute}}{\text{equivalent wt solute}}$$

Notice the similarity in the definitions of normality and equivalents to the definitions of molarity and moles. The only difference is that the (formula) weight and the equivalent weight of a solute are not necessarily the same. In acid-base reactions the equivalent weight of an acid or base is defined as the weight of the substance that donates (acid) or accepts (base) one mole of protons. In oxidation-reduction reactions the equivalent weight is the weight of a substance that donates (reducing agent) or accepts (oxidizing agent) one mole of electrons. For example, in the acid-base reaction

$$H_2SO_4 + 2\,NaOH \rightarrow Na_2SO_4 + 2\,H_2O$$

each mole of H_2SO_4 is donating *two* moles of protons, so the equivalent weight of H_2SO_4 is just the weight of $1/2$ mole of H_2SO_4, or $1/2 \times 98.0 = 49.0$ grams.

Similarly, in the oxidation-reduction reaction of metallic zinc with aqueous H^+,

$$Zn^{\circ}(s) + 2\,H^+(aq) \rightarrow Zn^{2+}(aq) + H_2(g)$$

one mole of zinc metal is giving up (donating) two moles of electrons. Therefore, the equivalent weight of zinc is the gram atomic weight divided by 2 ($65.4\,g \div 2 = 32.7\,g$).

The principal advantage to using normality and the concept of "equivalents" is that in any reaction the number of equivalents of one reactant must equal the number of equivalents of the other. In a titration, since equivalents = N $\times$ liters solution,

$$N_1 V_1 = N_2 V_2$$

where 1 and 2 are an acid and a base, or an oxidizing and a reducing agent, respectively, and V is the solution volume in liters (or milliliters).

Example 1.

Calcium hydroxide, $Ca(OH)_2$, is to be reacted with phosphoric acid, H_3PO_4, to give calcium phosphate, $Ca_3(PO_4)_2$, and water. Two solutions are prepared by adding 5.0 grams of $Ca(OH)_2$ to water to give 500 ml of solution and by adding 10.0 grams of H_3PO_4 to water to give 750 ml solution. Calculate the normality of the $Ca(OH)_2$ and H_3PO_4 solutions.

Solution.

One mole of $Ca(OH)_2$ accepts two moles of protons in the reaction; therefore, the equivalent weight of $Ca(OH)_2$ is $\frac{1}{2}$ the formula weight, or $\frac{1}{2} \times 74.1 = 37.1$. One mole of H_3PO_4 donates three moles of H^+ in the reaction, so the equivalent weight is $\frac{1}{3} \times 98.0 = 32.3$.

Thus,

$$\text{equivalents } Ca(OH)_2 = \frac{wt}{equiv\ wt} = \frac{5.0 \text{ grams}}{37.1 \text{ grams}} = 0.13 \text{ equivalent}$$

$$\text{equivalents } H_3PO_4 = \frac{wt}{equiv\ wt} = \frac{10.0 \text{ grams}}{32.3 \text{ grams}} = 0.31 \text{ equivalent}$$

From the definition of normality:

$$N\ (Ca(OH)_2) = \frac{0.13 \text{ equivalent}}{0.500 \text{ liter solution}} = 0.26 \text{ equivalent per liter}$$

and,

$$N\ (H_3PO_4) = \frac{0.31 \text{ equivalent}}{0.750 \text{ liter solution}} = 0.41 \text{ equivalent per liter}$$

Example 2.

A 0.10 N sodium thiosulfate, $Na_2S_2O_3$, solution is used to titrate iodine, I_2, in an iodine solution of unknown concentration. The products of the reaction are $Na_2S_4O_6$ and I^-. Starch indicator shows the reaction is complete when 36.2 ml of the thiosulfate solution has been added to 25.0 ml of the iodine solution. Find the normality of the I_2 solution.

Solution.

Since the number of electrons lost by $S_2O_3^{2-}$ ions must equal the number of electrons gained by I_2,

$$\text{equivalents } S_2O_3^{2-} = \text{equivalents } I_2$$

or,

$$N \times V\ (S_2O_3^{2-}) = N \times V\ (I_2)$$

and substituting in the equation

$$(0.10 \text{ N}) \times (36.2 \text{ ml}) = N \times (25.0 \text{ ml}), N\ (I_2) = 0.145$$

Why is it permissible to use milliliters directly without converting volume to liters?

Mole Fraction

Mole fraction, X, is defined as

$$X = \frac{\text{moles of one component}}{\text{total moles of all components}}$$

Mole fraction is useful in describing concentrations in multi-component systems and is often the concentration term used in studies of partial pressures of gases and vapor pressures of components of solutions.

Example 3.

A solution to be used as a hand lotion is prepared by mixing 90.0 grams of water, 6.40 grams of methyl alcohol (CH_3OH), and 18.40 grams of glycerol ($C_3H_8O_3$). Calculate the mole fraction of glycerol.

Solution.

The number of moles of each component is found from, moles = wt/MW:

$$\text{moles } H_2O = \frac{90.0}{18.0} = 5.0$$

$$\text{moles } CH_3OH = \frac{6.40}{32.0} = 0.20$$

$$\text{moles } C_3H_8O_3 = \frac{18.40}{92.0} = 0.20$$

So, the mole fraction of glycerol is

$$X = \frac{0.20}{5.0 + 0.20 + 0.20} = 0.037$$

CATEGORY V   SOME EFFECTS OF CONCENTRATION ON SOLUTION PROPERTIES

If we examine a solution from the solvent's point of view, it's a rather frustrating experience. All of those solute particles tend to get in the way of the solvent molecules when the solvent tries to behave like it should. For one thing, the tendency for solvent molecules to "escape" from the liquid phase to the vapor phase is partially blocked, i.e., the vapor pressure is lowered. This effect is described quantitatively by Raoult's Law, which states:

$$P_1 = X_1 P_1^0$$

where P_1 is the vapor pressure of a solvent in a solution, X_1 is the mole fraction of the solvent, and P_1^0 is the vapor pressure of the pure solvent. Raoult's Law is an example of what is called a *colligative property*, which means that it depends strictly upon the number (concentration) of solute particles and not upon the nature of the solute particles. Thus, as far as the solvent is concerned, one mole of sugar molecules would have exactly the same effect as one mole of alcohol molecules for a given amount of solvent.

The *boiling point* of a substance is defined as that temperature at which the vapor pressure of the substance is equal to the external pressure. Obviously, when the vapor pressure of a substance is lowered, it will require a higher temperature to reach the boiling point. The boiling points of solutions are higher than the boiling points of pure solvents. This effect may be described by the relationship;

$$\Delta T_b = B \times m$$

where ΔT_b is the change in boiling point, B is a constant specific for a particular solvent, and m is the molality of *solute particles* in the solution. Some typical values for B, the boiling point constant, are water, $0.52°C$; ethanol, $1.22°C$; benzene, $2.53°C$; carbon tetrachloride, $5.03°C$.

Another instance in which a solvent (in a solution) is frustrated in its attempt to behave normally is when it begins to freeze. The solute particles get in the way of establishing an orderly solvent crystal structure and a lower temperature is required before freezing can occur. The freezing (or melting) point of a solution is lowered compared to a pure solvent. This effect is quantitatively defined by the equation

$$\Delta T_f = F \times m$$

where ΔT_f is the change in freezing point, F is the freezing point constant for the solvent, and m is the molality of solute particles in the solution. Some typical values of F are water, $1.86°C$; benzene, $5.10°C$; naphthalene, $6.9°C$; and camphor, $40.0°C$.

Let's look at some problems that illustrate these different, but related, effects of solution concentration.

Example 1.

A solution, to be used as an "antifreeze" in an automobile, is prepared by dissolving 100 g of ethylene glycol, $C_2H_6O_2$, per 200 g of water.
  (a) The vapor pressure of pure water is 23.8 mm Hg at $25°C$. Find the vapor pressure of water for the solution at $25°C$.
  (b) Find the boiling point of the solution at 1 atm pressure.
  (c) Find the freezing point of the solution.

Solution.

  (a) Since the molecular weight of $C_2H_6O_2$ is 62.0, the number of moles of $C_2H_6O_2$ in the solution = 100 g/62.0 g/mole = 1.61. Similarly, we have (200 g ÷ 18.0) = 11.11 moles of H_2O. The mole fraction of H_2O is

$$X_1 = \frac{11.11 \text{ moles}}{11.11 \text{ moles} + 1.61 \text{ moles}} = 0.873$$

Substituting into Raoult's Law, we find

$$P_1 = X_1 P_1^0 = 0.873 \times 23.8 \text{ mm} = 20.8 \text{ mm Hg}$$

  (b) The molality of the non-electrolyte $C_2H_6O_2$ in the solution is

$$m = \frac{1.61 \text{ moles } C_2H_6O_2}{0.200 \text{ kg } H_2O} = 8.05 \text{ moles per kg}$$

The boiling point of pure water (1.0 atm pressure) is $100.0°C$ and the boiling point constant of water is $0.52°C$. The change in the boiling point is

$$\Delta T_b = B \times m = 0.52 \times 8.05 = 4.19°C$$

The boiling point of the solution is $104.19°C$.
  (c) The freezing point of pure water at 1.0 atm is $0°C$. The freezing point

constant for water is $1.86°C$. Compared to pure water, the freezing point of the solution is decreased by

$$\Delta T_f = F \times m = 1.86 \times 8.05 = 14.97°C$$

Therefore, the freezing point of the solution is $-14.97°C$.
If we translate that into the, as yet, more familiar $°F$ ($°F = 32 + 1.8°C$), the freezing point of the antifreeze solution is $+5°F$. Most commercial "antifreeze" is just ethylene glycol. It is easy to see why people with water-cooled automobiles who live in cold climates must use the stuff by the gallon!

Example 2.

Find the freezing point of an aqueous 0.20 m NaCl solution.

Solution.

Colligative properties depend upon the total *number* of solute particles, not the kind of particles. Na^+ ions and Cl^- ions have an identical effect upon the freezing point of water. Since NaCl is a strong electrolyte (ionizes completely), a 0.20 m NaCl solution is actually 0.40 molal in solute particles. The freezing point of the solution, compared to pure water, should be lowered by

$$\Delta T_f = F \times m = 1.86 \times 0.40 = 0.74°C$$

The solution should freeze at $-0.74°C$ if NaCl were 100% ionized. The experimentally observed freezing point of a 0.20 m NaCl solution is $-0.69°C$, indicating that even NaCl, one of the strongest electrolytes, is not 100% ionized at a concentration of 0.20 m. Ionization is 100% at a concentration of 1.0×10^{-5} m. Why is salt sprinkled on streets and sidewalks in the winter? Why isn't salt used in car radiators?

Example 3.

The standard freezing point of benzene is $5.50°C$ and the freezing point constant is $5.10°C$. When 2.00 g of an organic non-electrolyte is dissolved in 50.0 g of benzene, the freezing point of the solution is $1.71°C$. Find the molecular weight of the organic substance.

Solution.

Freezing and boiling points of solutions can be used to determine molecular weights of unknown materials. In this example, the freezing point is lowered from $5.50°C$ to $1.71°C$, hence, $\Delta T_f = 3.79°C$. We can find the molality of the solute:

$$\Delta T_f = F \times m; \quad m = \frac{\Delta T_f}{F} = \frac{3.79°C}{5.10°C/m} = 0.743 \text{ m}$$

Knowing the molality of the unknown, U, we can easily find the MW. If there are 0.743 mole U per 1000 g benzene, then there must be $0.743 \div 20 = 0.0372$ mole in 50.0 g benzene. Then, 2.00 g is 0.0372 mole and MW = Wt $\div$ no. moles = $2.00 \div 0.0372$ = 53.8 g/mole.

PROBLEMS   *Category I*

1. If 0.25 mole of KBr is dissolved in enough water to give 250 ml of solution, what is the molarity of the solution?

2. What is the molarity of a solution prepared by dissolving 9.20 grams of formic acid, HCOOH, in water to give 1500 ml of solution?

3. Twenty grams of methyl alcohol, CH_3OH, and 80.0 grams of water are mixed. The density of the resulting solution is 0.970 gram per ml. Calculate the molarity of the solution.

4. If 10.0 ml of 18.0 M HNO_3 is diluted to a volume of 200 ml, what is the molarity of the dilute solution?

Category II

5. How many moles of H_2SO_4 are there in 500 ml of 1.50 M H_2SO_4 solution?

6. How many moles of iron(III) nitrate, $Fe(NO_3)_3$, would be required to prepare 100 ml of 0.30 M $Fe(NO_3)_3$ solution?

7. How many grams of glucose, $C_6H_{12}O_6$, are there in 150 ml of 0.50 M glucose solution?

8. How many grams of KOH would be needed to prepare 250 ml of 0.10 M KOH solution?

9. Twenty ml of a 1.0 M solution of the amino acid alanine, $C_3H_7O_2N$, is diluted to a volume of 200 ml. How many grams of alanine would be contained in 40.0 ml of the dilute solution?

Category III

10. What volume of 0.15 M $AgNO_3$ solution would be required to give 0.002 mole of Ag^+?

11. How many mls of a solution in which $Cl^- = 0.40$ M would be needed to yield 1.5×10^{-4} mole of chloride ion?

12. When tetraethyl lead, $Pb(C_2H_5)_4$, is added to gasoline as an "antiknock" agent, the solution is about 1.0×10^{-6} M in $Pb(C_2H_5)_4$. How many liters of gasoline would be required to give 1.0 gram of lead?

13. Fluoridated water contains "one part per million, ppm," of F^- ion by weight. Convert this to molarity and calculate the number of liters of water required to give 0.01 mole of F^-. (Density of water = 1.00 g/ml.)

Category IV

14. If 3.55 grams of DDT, $C_{14}H_9Cl_5$, is dissolved in 500 grams of benzene, C_6H_6, calculate the molality of the solution.

15. How many grams of acetone, C_3H_6O, would have to be added to 250 grams of water to prepare a 0.50 m (molal) solution?

16. A sulfurous acid, H_2SO_3, solution is prepared by dissolving 16.4 grams of H_2SO_3 in enough water to give 500 ml of solution. Calculate the molarity of the solution. The H_2SO_3 solution is used in the following reactions:

$$H_2SO_3 + 2\,KOH \rightleftarrows K_2SO_3 + 2\,H_2O$$

$$H_2SO_3 + Br_2 + H_2O \rightleftarrows 4\,H^+ + SO_4{}^{2-} + 2\,Br^-$$

What is the equivalent weight of H_2SO_3 as an acid? As a reducing agent? What is the normality of the H_2SO_3 solution as an acid? As a reducing agent?

17. A solution of an unknown acid is titrated with 0.20 N NaOH. The acid solution was prepared by dissolving 3.00 grams of the acid in enough water to give 50.0 ml solution. The acid solution required 35.5 ml of the NaOH solution for the neutralization reaction to be complete. What is the normality of the unknown acid solution? What is the equivalent weight of the unknown acid?

18. Analysis of a sample of the rocket exhaust gas of the Viking spacecraft showed that it contained 24.3 grams H_2O, 17.6 grams CO_2, 8.70 grams NO_2 and 2.35 grams NO. Calculate the mole fraction of NO_2 in the sample.

Category V

19. Find the vapor pressure of water above a solution containing 80.0 g of ethyl alcohol, C_2H_5OH, in 100.0 g of water (at $25°C$).

20. The standard boiling point of benzene is $80.0°C$. Calculate the boiling point of a solution prepared by dissolving 355.0 g of DDT, $C_{14}H_9Cl_5$, in 500.0 g of benzene.

21. The Great Salt Lake in Utah contains 12% sodium chloride by weight. Find the freezing point of the lake.

22. When 2.60 g of a molecular unknown was dissolved in 50.0 g of water, the freezing point of the solution was found to be $-1.61°C$. Calculate the molecular weight of the unknown.

23. The ionization of hydrofluoric acid, $HF \rightleftarrows H^+ + F^-$, is not complete in aqueous solution. When 0.100 mole of HF is dissolved in 100.0 g of water, the freezing point of the solution is $-1.91°C$. How much of the HF is ionized?

General Problems

24. Complete the following table.

Solute	Moles Solute	Grams Solute	Volume Solution	Molarity
HCl	0.15		0.30 ℓ	
CH_3COOH		12.0		0.40
NH_3	0.80			0.20
$C_{12}H_{22}O_{11}$		6.84	100 ml	

25. How many mls of 0.008 M HNO_3 solution could be prepared from 1.0 ml of 16 M HNO_3?

26. The label on a bottle of aqueous sulfuric acid, H_2SO_4, gives its concentration as 44.0% by weight, and the density of the solution is listed as 1.31 g/ml. Convert this information to
(a) molarity   (b) molality   (c) mole fraction

27. When 34.0 g of silver nitrate, $AgNO_3$, is added to 100.0 ml of water (density 1.00 g/ml), the final volume of the solution is 110.0 ml. Find
(a) the molarity of the $AgNO_3$ (b) the molality of the $AgNO_3$
(c) the density of the solution in g/ml

28. When 10.2 g of the strong electrolyte $Ca(NO_3)_2$ is added to 1.00 liter of water, the density of the solution is 1.01 g/ml. Calculate
(a) the molarity of $Ca(NO_3)_2$ (b) the molarity of NO_3^-
(c) the molality of $Ca(NO_3)_2$ (d) the total molality of all solute species
(e) the freezing point of the solution (assume 100% ionization)

29. When 10.0 g of a non-electrolyte were dissolved in 100.0 g of water, the resulting solution had a freezing point of $-0.614°C$ and its density was found to be 1.08 g/ml. Find the *molarity* of the solute in the solution.

30. A solution is prepared by dissolving 15.6 g benzene, C_6H_6, in 500 g of carbon tetrachloride, CCl_4. The volume of the solution is 450 ml. Find
(a) the molality   (b) the molarity   (c) the density
(d) the boiling point of the solution

8 CHEMICAL EQUILIBRIUM (GASES)

Equilibrium may be the most important concept in the science of chemistry. In principle most (and perhaps all) chemical reactions are reversible, and reversible reactions are best understood in terms of the concept of chemical equilibrium. Imagine some general reaction A → B. The rate of the reaction depends upon the amount (concentration) of A available. If the reaction is reversible, this means that the reaction A ← B will also occur, and the rate of that reaction will depend upon the concentration of B available. As the forward reaction A → B proceeds, the concentration of A decreases and the concentration of B increases. As the concentration of A decreases, so does the rate of the forward reaction, and as the concentration of B increases, so does the rate of the reverse reaction. Eventually the rates of the forward and reverse reactions become equal and from that point on, the system appears to undergo no further change. That is the point of *equilibrium.*

$$qA + rB \rightleftarrows sC + tD$$

the equilibrium constant is

$$\frac{[C]^s \, [D]^t}{[A]^q \, [B]^r} = K$$

The square brackets indicate equilibrium concentrations in moles per liter. Specific examples are

$$N_2 + 3\,H_2 \rightleftarrows 2\,NH_3 \, ; \quad \frac{[NH_3]^2}{[N_2] \, [H_2]^3} = K$$

$$2\,CO_2 \rightleftarrows 2\,CO + O_2 \, ; \quad \frac{[CO]^2 \, [O_2]}{[CO_2]^2} = K$$

While concentrations of reactants or products can change, the equilibrium constant must be constant (at a given temperature). Thus, any reversible reaction will proceed until the concentration quotient is equal to K. Moreover, any changes in concentrations of reactants or products in a system already at equilibrium will cause either forward or reverse reaction to occur until the concentration quotient is again equal to K (Le Chatelier's Principle). With this in mind, we will divide general equilibrium problems into two categories.

Problem Categories

I   Attaining equilibrium
II Changes at equilibrium

CATEGORY   ATTAINING EQUILIBRIUM
I

Example 1.

A mixture of $CO(g)$ and $Cl_2(g)$ are placed in a one liter container where they react (reversibly) to form $COCl_2(g)$. When equilibrium is reached, analysis shows there are 0.30 mole of CO, 0.20 mole of Cl_2, and 0.80 mole of $COCl_2$ in the container. Calculate the equilibrium constant for the reaction

$$CO(g) + Cl_2(g) \rightleftarrows COCl_2(g)$$

Solution.

We need only substitute the appropriate numbers into the equilibrium constant expression:

$$\frac{[COCl_2]}{[CO]\ [Cl_2]} = K$$

Since the volume is, conveniently, one liter, the number of moles and the molarity of each constituent are the same. Substituting we find

$$\frac{(0.80)}{(0.30)\ (0.20)} = 13.3 = K$$

Example 2.

It is known that sulfur dioxide and oxygen react reversibly to give sulfur trioxide:

$$2\ SO_2(g) + O_2(g) \rightleftarrows 2\ SO_3(g)$$

Exactly 0.80 mole of SO_2 and 0.60 mole of O_2 are mixed in a 1.0 liter vessel, and when equilibrium is reached, 0.60 mole of SO_3 are found in the container. Calculate K for the reaction.

Solution.

Examine the equation for the reaction. We see that every 2 moles of SO_3 formed requires 2 moles of SO_2 and one mole of O_2. So, if 0.60 mole of SO_3 were formed by the reaction, it must be that 0.60 mole of SO_2 and 0.30 mole of O_2 were "used up" in reaching equilibrium. Since we started with 0.80 mole of SO_2, there would only be 0.20 mole of SO_2 left at equilibrium. Similarly, we started with 0.60 mole of O_2, but 0.30 mole of O_2 reacted, so at equilibrium there will only be 0.30 mole of unreacted O_2 left in the container. Now we have the numbers we need to plug into the equilibrium constant expression:

$$\frac{[SO_3]^2}{[SO_2]^2\ [O_2]} = K = \frac{(0.60)^2}{(0.20)^2\ (0.30)} = 30.0$$

Note that, as in Example 1, the volume of the container is one liter, so the number of moles is the same as the concentration in moles per liter.

Example 3.

The equilibrium constant for the reaction

$$SO_2 (g) + NO_2 (g) \rightleftarrows SO_3 (g) + NO (g)$$

is 3.0 at 400°C. If 1.0 mole of SO_2 and 1.0 mole of NO_2 are injected into a 500 ml flask at 400°C, how many moles of SO_3 and NO will there be in the flask at equilibrium?

Solution.

This problem poses many problems, but if you look at it one step at a time, it's easy. First, of course, we must see that the volume of the container is not one liter, and the initial concentrations of SO_2 and NO_2 are not 1.0 M but 1.0 mole/0.50 liter = 2.0 M. We know that SO_2 and NO_2 will react until at equilibrium

$$\frac{[SO_3] \ [NO]}{[SO_2] \ [NO_2]} = 3.0$$

Since the ratio of products to reactants is greater than 1.0 at equilibrium, it must be that more than half of the initial 2.0 moles per liter of SO_2 and NO_2 are consumed in the reaction. We could go through a tedious process of guesses until the ratio came out exactly 3.0. For example, we might guess that 1.5 moles per liter of the SO_2 and NO_2 are used up. At equilibrium, that would give 1.5 moles per liter of SO_3 and NO, and only 0.50 mole per liter of SO_2 and NO_2 left. We could then plug in those values and see if the ratio was 3.0. If that didn't work, then we could guess again, and so on, until the numbers came out right. Instead, let's apply some elementary algebra.

Choose an unknown, let's say $[SO_3]$, and call it "X." The balanced equation for the reaction tells us that for every mole (per liter) of SO_2 and NO_2 that react, an identical amount of SO_3 and NO must be formed. So, if $[SO_3] = X$, so too does $[NO] = X$. If X moles per liter of SO_3 and NO are formed, then X moles per liter of SO_2 and NO_2 must be used up. At equilibrium, the amount of SO_2 and NO_2 remaining will be what we started with minus the amount used up, i.e., $[SO_2] = 2.0 - X$, and $[NO_2] = 2.0 - X$.

$$\begin{array}{ccccccc} (2.0-X) & & (2.0-X) & & X & & X \\ SO_2 & + & NO_2 & \rightleftarrows & SO_3 & + & NO \end{array}$$

Substituting into the equilibrium constant, we now have an equation with only one unknown, X:

$$\frac{[SO_3] \ [NO]}{[SO_2] \ [NO_2]} = \frac{(X) \ (X)}{(2.0-X) \ (2.0-X)} = 3.0$$

Often, in this type of equilibrium problem, one ends up with a quadratic equation. In this case, a quadratic can be conveniently avoided by taking the square root of both sides of the equation, leaving

$$\frac{(X)}{(2.0-X)} = 1.73$$

Solving we get:

$$X = 3.46 - 1.73X$$

$$2.73X = 3.46$$

$$X = 1.27 \ (\simeq 1.3)$$

Thus, at equilibrium, there will be 1.27 moles per liter of SO_3 and 1.27 moles per liter of NO in the flask. How many moles per liter of SO_2 and NO_2 are left? Remember, the flask has a volume of 500 ml, so the number of moles of SO_3 and NO in the flask will be 1.27 M $\times$ 0.50 ℓ = 0.635 mole.

Example 4.

Dinitrogen tetroxide, N_2O_4, reacts to give nitrogen dioxide, NO_2, according to the equilibrium N_2O_4 (g) $\rightleftarrows$ 2 NO_2 (g). At 25°C the equilibrium constant for the reaction is K = 5.0 $\times$ 10^{-3}. If one mole of N_2O_4 is placed in a 1.0 liter container, how many moles of NO_2 will there be at equilibrium?

Solution.

When solving any equilibrium problem, it is a good idea to begin by writing down the equation for the reaction and the equilibrium constant expression:

$$N_2O_4 \rightleftarrows 2\,NO_2$$

$$\frac{[NO_2]^2}{[N_2O_4]} = 5.0 \times 10^{-3}$$

Then, ask yourself, "What is the unknown?" In this problem we need the moles, or moles per liter, of NO_2. Let's call $[NO_2]$ = X. Notice that two NO_2 molecules are formed from each N_2O_4 molecule that reacts, so, if X moles per liter of NO_2 are formed, the amount of N_2O_4 used up must be ½ X. If we had elected to define our unknown, X, as the moles per liter of N_2O_4 used up, what would $[NO_2]$ be?

So, at equilibrium we can set $[NO_2]$ = X, and $[N_2O_4]$ = initial concentration − concentration reacted = 1.0 − ½ X. We can now substitute into the equilibrium constant expression and solve for the one unknown, X:

$$\frac{(X)^2}{(1.0 - 0.5X)} = 5.0 \times 10^{-3}$$

$$X^2 = 5.0 \times 10^{-3} - 2.5 \times 10^{-3}\,X$$

or

$$X^2 + 2.5 \times 10^{-3}\,X - 5.0 \times 10^{-3} = 0$$

We have the familiar quadratic equation detested by algebra students everywhere:

$$aX^2 + bX + c = 0$$

and we can solve for X by applying the quadratic equation

$$X = \frac{-b \pm \sqrt{b^2 - 4ac}}{2a}$$

$$X = \frac{-2.5 \times 10^{-3} \pm \sqrt{6.25 \times 10^{-6} - (4)(1)(-5.0 \times 10^{-3})}}{2 \times 1}$$

$$X = \frac{-2.5 \times 10^{-3} + \sqrt{2.0 \times 10^{-2}}}{2} \quad \text{(the negative root would not apply)}$$

$$X = 0.069$$

An easier way of solving the arithmetic in this problem involves judicious use of an approximation. Examine the original set up:

$$\frac{(X)^2}{(1.0 - 0.5X)} = 5.0 \times 10^{-3}$$

For the ratio $(X)^2/(1.0 - 0.5X)$ to be only 0.005/1, it must be that X is small. If X is small enough, then subtracting 0.50X from 1.0 in the denominator will not have much of an effect, i.e., the denominator will still be approximately 1.0. Let's make that assumption and see what happens. We will neglect 0.5X in the denominator. Then we have

$$\frac{(X)^2}{1.0} = 5.0 \times 10^{-3}$$

$$X = \sqrt{1.0 \times 5.0 \times 10^{-3}}$$

$$X = 0.071$$

Not bad! Depending upon the accuracy required, this type of approximation can and should be used when solving equilibrium problems with large or small equilibrium constants. If it doesn't work, then one can always resort to the quadratic formula.

CHANGES AT EQUILIBRIUM   CATEGORY II

The previous examples have all dealt with systems at, or attaining, equilibrium. Now let us examine some problems that involve changing systems at equilibrium. Remember, the only thing that can change the value of an equilibrium constant is a change in temperature, but concentrations of reactants or products can be changed simply by adding or taking out some of the appropriate material, or by changing the volume.

Example 1.

At a certain temperature, K = 0.051 for the reaction

$$CO(g) + H_2O(g) \rightleftarrows CO_2(g) + H_2(g)$$

In an experiment, the system is at equilibrium with $[CO_2] = 0.90$, $[H_2] = 0.90$, $[CO] = 4.0$, and $[H_2O] = 4.0$. An additional 1.0 mole per liter of CO and H_2O are added to the container. Find the concentrations of CO, H_2O, CO_2 and H_2 when the system again reaches equilibrium.

Solution.

The equilibrium constant was satisfied by the original concentrations:

$$\frac{[CO_2]\,[H_2]}{[CO]\,[H_2O]} = \frac{(0.90)\,(0.90)}{(4.0)\,(4.0)} = 0.051$$

So, K must equal 0.051 for the system. But, by adding an additional mole per liter of CO and H_2O, we make $[CO] = 5.0$, and $[H_2O] = 5.0$, and the equilibrium expression is no longer equal to the equilibrium constant:

$$\frac{(0.90)\,(0.90)}{(5.0)\,(5.0)} = 0.032 \neq 0.051$$

So, something has to happen. In order for the equilibrium expression (ratio) to again become equal to 0.051, the numerator must get larger, the denominator get smaller, or both. This is accomplished by "shifting" the equilibrium to the right, i.e., some CO and H_2O will react to form CO_2 and H_2O until the ratio is again 0.051. How much? If we call the concentration of CO_2 formed "X," then the concentration of H_2 formed must also be X, and the concentration of CO and H_2O used up would also be X. Thus, we can define all of the new concentrations at equilibrium in terms of the one unknown:

$$\begin{array}{ccccccc} (5.0-X) & & (5.0-X) & & (0.90+X) & & (0.90+X) \\ CO & + & H_2O & \rightleftarrows & CO_2 & + & H_2 \end{array}$$

$$\frac{[CO_2]\,[H_2]}{[CO]\,[H_2O]} = \frac{(0.90+X)\,(0.90+X)}{(5.0-X)\,(5.0-X)} = 0.051$$

We can simplify the math by taking the square root of both sides. Then, solve for X.

$$\frac{(0.90+X)}{(5.0-X)} = 0.23; \; 0.90 + X = 1.15 - 0.23X$$

$$1.23X = 0.25; \; X = 0.20$$

So, when equilibrium is re-established, $[CO_2] = [H_2] = 0.90 + 0.20 = 1.10$, and $[CO] = [H_2O] = 5.0 - 0.20 = 4.8$.

Example 2.

The reaction $CO(g) + Cl_2(g) \rightleftarrows COCl_2(g)$ is at equilibrium with $[CO] = 0.30$, $[Cl_2] = 0.20$, and $[COCl_2] = 0.80$. How many moles (per liter) of $COCl_2$ must be added in order to increase the concentration of Cl_2 to 0.30 M? K = 13.3.

Solution.

Adding $COCl_2$ will shift the equilibrium to the left and increase the amounts of Cl_2 and CO while consuming some $COCl_2$. Examine the equilibrium

$$CO = Cl_2 \rightleftarrows COCl_2$$

For every Cl_2 molecule produced, a molecule of CO must also be produced. So, increasing the moles per liter of Cl_2 by 0.10 (from 0.20 to 0.30) must also increase the concentration of CO by 0.10, from 0.30 to 0.40. Since we know that the concentrations of Cl_2 and CO must be 0.30 and 0.40 when equilibrium is re-established, we have only to find the new concentration of $COCl_2$. We could define the unknown as the total concentration of $COCl_2$, $[COCl_2] = X$, or as the amount to be added, in which case at equilibrium $[COCl_2]$ would be (0.80 + X − 0.10). Let's solve both ways. First, for total $COCl_2$,

$$\frac{[COCl_2]}{[CO]\,[Cl_2]} = \frac{(X)}{(0.40)\,(0.30)} = 13.3$$

$$X = 1.60$$

If the total $[COCl_2]$ must be 1.60 M, then obviously 0.90 moles per liter must be added to the 0.80 moles per liter originally in the container. However, this may not seem so obvious. Notice that what we have solved for is the new concentration of $COCl_2$ at equilibrium, 1.60 M. But, in reaching that equilibrium *0.10* mole per liter of the $COCl_2$ was *used up* to form CO and Cl_2, so a total of 0.90 mole per liter must be added to the 0.80 mole per liter originally present in order to give *1.70 M* – 0.10 M = *1.60 M.*

That is why, if we define our unknown as the amount added, the equilibrium concentration of $COCl_2$ is (0.80 + X – *0.10*). If we do that, and solve for X, we find

$$\frac{(0.80 + X - 0.10)}{(0.40)\,(0.30)} = 13.3; \quad X = 0.90$$

Example 3.

The reaction H_2 (g) + I_2 (g) $\rightleftarrows$ 2 HI has an equilibrium constant of 6.0 at a certain temperature. If the system is equilibrated at $[H_2]$ = 0.50, $[I_2]$ = 0.50, and [HI] = 1.23 in a one liter flask, and suddenly 0.60 mole of HI are removed, what will be the concentrations of all species when equilibrium is re-established?

Solution.

$$H_2 + I_2 \rightleftarrows 2\,HI$$

$$\frac{[HI]^2}{[H_2]\,[I_2]} = 6.0$$

Removal of some HI will shift the equilibrium to the right, using up some H_2 and I_2 to form HI until equilibrium is reached. Immediately upon removal of 0.60 mole of HI, we are left with $[H_2]$ = $[I_2]$ = 0.50, and [HI] = 0.63. We could define the unknown as amount of HI to be formed, or amount of H_2 (or I_2) used up. Let's set X = moles per liter of H_2 that react. Then, when equilibrium is reached, $[H_2]$ = (0.50 – X). One I_2 reacts for every H_2 that reacts, so we will also have $[I_2]$ = (0.50 – X). For each H_2 or I_2 used there are *two* HI formed, so at equilibrium [HI] = (0.63 + 2X). Now we can plug into the equilibrium constant expression, take the square root of both sides, and solve for X.

$$\frac{(0.63 + 2X)^2}{(0.50 - X)\,(0.50 - X)} = 6.0; \quad \frac{(0.63 + 2X)}{(0.50 - X)} = 2.45$$

$$0.63 + 2X = 1.23 - 2.45X; \quad 4.45X = 0.60$$

$$X = 0.135 \simeq 0.14$$

So, when equilibrium is re-established, $[H_2]$ = $[I_2]$ = (0.50 – 0.14) = 0.36, and [HI] = (0.63 + 2 × 0.14) = 0.91.

Example 4.

The reaction

$$S_2Cl_4 \text{ (g)} \rightleftarrows 2\,SCl_2 \text{ (g)}$$

has an equilibrium constant of 25.0 at a certain temperature. If 5.0 moles of SCl_2 are in equilibrium with 1.0 mole of S_2Cl_4 in a 1.0 liter container, what will happen if the volume of the container is decreased to 0.50 liter? Find the concentrations of SCl_2 and S_2Cl_4 at the 0.50 liter volume.

Solution.

Cutting the volume by 2 will double the molar concentrations of SCl_2 and S_2Cl_4 and the system will no longer be at equilibrium:

$$\frac{(5.0)^2}{1.0} = \frac{[SCl_2]^2}{[S_2Cl_4]} = 25; \text{ but } \frac{(10.0)^2}{2.0} = 50$$

so the equilibrium will shift to the left, forming more S_2Cl_4 and consuming SCl_2 until the ratio $[SCl_2]^2/[S_2Cl_4]$ is again equal to 25.0.

If we define the moles per liter of S_2Cl_4 formed as X, then the moles per liter of SCl_2 consumed must be 2X. At equilibrium we would find

$$\frac{[SCl_2]^2}{[S_2Cl_4]} = \frac{(10.0 - 2X)^2}{(2.0 + X)} = 25.0$$

$$100 - 40X + 4X^2 = 50 + 25X$$

$$4X^2 - 65X + 50 = 0$$

and solving by the quadratic formula we find

$$X = 15.4 \text{ or } 0.81$$

The correct root must be 0.81 and at equilibrium $[SCl_2] = 10.0 - (2 \times 0.81) = 8.4$ M, and $[S_2Cl_4] = 2.0 + 0.81 = 2.8$ M.

Thus we see that in a gas phase equilibrium if there are an unequal number of moles on the reactant and product side of the equilibrium, changing volume will cause a shift in the equilibrium. One other factor that will change an equilibrium is temperature. Equilibrium constants and free energy changes are related by the equation;

$$\Delta G° = -RT \ln K$$

where $\Delta G°$ is the standard free energy change, R is the gas constant, T is the temperature in $°K$, and K is the equilibrium constant. The magnitude of K depends upon T. In general, for an endothermic reaction, increasing T will increase K and for an exothermic reaction, an increase in T will decrease K.

PROBLEMS   *Category I*

1. One mole of PCl_5 was placed in a 1.0 liter vessel. At equilibrium, 0.20 mole of Cl_2 were found in the vessel. Find the equilibrium constant for the reaction $PCl_5(g) \rightleftarrows PCl_3(g) + Cl_2(g)$.

2. Exactly 5.60 g of carbon monoxide, CO, and 3.20 g of O_2 are mixed in a 200 ml flask. At equilibrium, 6.60 g of carbon dioxide, CO_2, are found in the flask. Calculate K for the reaction $2 CO(g) + O_2(g) \rightleftarrows 2 CO_2(g)$.

3. The equilibrium constant for the reaction $2 HI(g) \rightleftarrows H_2(g) + I_2(g)$ is 0.016 at 500°C. If 2.0 moles of HI are placed in a 1.0 liter container, calculate $[H_2]$ and $[I_2]$ at equilibrium.

4. For the reaction $2 HBr(g) \rightleftarrows H_2(g) + Br_2(g)$, $K = 2.5 \times 10^{-19}$ at 25°C. If 2.0 moles of HBr are placed in a 1.0 liter container, find $[H_2]$ and $[Br_2]$ at equilibrium.

5. Exactly 3.0 moles of S_2Cl_4 are introduced into a 5.0 liter container. The reaction $S_2Cl_4(g) \rightleftarrows 2 SCl_2(g)$ occurs, and at equilibrium only 0.20 mole of S_2Cl_4 remain in the container. Calculate K for the reaction.

6. For the reaction $SO_2(g) + NO_2(g) \rightleftarrows SO_3(g) + NO(g)$ the equilibrium constant is 3.0 at 27°C. If 1.0 mole of SO_2 and 2.0 moles of NO_2 are placed in a 1.0 liter container, find the concentrations of all species at equilibrium.

7. If $K = 1.8$ for the reaction $N_2(g) + 3 H_2(g) \rightleftarrows 2 NH_3(g)$, how many moles of NH_3 must be placed in a 1.0 liter container in order to yield an equilibrium concentration for H_2 of 6.0 M?

Category II

8. Consider the equilibrium $2 NO(g) + 2 CO(g) \rightleftarrows 2 CO_2(g) + N_2(g)$. If equilibrium is established, which way will it shift if
(a) CO_2 is added?　　　　(b) N_2 is removed?　　　　(c) NO is added?
(d) the volume of the container is suddenly decreased?

9. For the reaction $2 SO_2(g) + O_2(g) \rightleftarrows 2 SO_3(g)$, $K = 5.0 \times 10^3$ at 500°C. The system is at equilibrium with $[SO_2] = 1.0$, $[SO_3] = 10.0$, and $[O_2] = 0.02$. How many moles per liter of SO_3 must be added in order that $[O_2] = 1.02$ at equilibrium?

10. An equilibrium mixture $XeF_2(g) + OF_2(g) \rightleftarrows XeOF_2(g) + F_2(g)$ was found to contain 0.60 mole of XeF_2, 0.30 mole of OF_2, 0.10 mole of $XeOF_2$ and 0.40 mole of F_2 in a one liter container. How many moles of OF_2 must be added to increase $[XeOF_2]$ to 0.20 M?

11. The equilibrium constant for the reaction $C_2H_2(g) + H_2(g) \rightleftarrows C_2H_4(g)$ is 4.0 at a certain temperature. The system is at equilibrium with $[C_2H_2[= 1.0$, $[H_2] = 1.0$, and $[C_2H_4] = 4.0$. Suddenly the concentration of C_2H_4 is decreased to 2.0 M by removing C_2H_4 from the container. Calculate the concentrations of all species when equilibrium is re-established.

12. For the reaction $H_2(g) + I_2(g) \rightleftarrows 2 HI(g)$, $K = 6.0$ at a certain temperature. The system is at equilibrium with $[H_2] = 2.0$, $[I_2] = 2.0$, and $[HI] = 4.9$. How many moles per liter of H_2 must be removed to increase $[I_2]$ to 2.5 M?

General Problems

13. For the equilibrium $N_2O_3(g) \rightleftarrows N_2O(g) + O_2(g)$, $K = 4.00$ at a certain temperature. Complete the following table.

	Original Concentration	Change in Concentration	Equilibrium Concentration
$[N_2O_3]$	0.50 M		
$[N_2O]$	0		
$[O_2]$	0.50 M		

14. Consider the following system at equilibrium:

$$22 \text{ kcal} + CH_4(g) + 2 H_2S(g) \rightleftarrows CS_2(g) + 4 H_2(g)$$

How will the equilibrium shift if
(a) H_2 is removed?   (b) H_2S is removed?   (c) the volume is doubled?
(d) the temperature is increased?

15. For the reaction $Cl_2(g) \rightleftarrows 2 Cl(g)$, $K = 3.6 \times 10^{-2}$ at 2000°C. How many moles of Cl_2 must be placed in a 3.0 liter container in order to yield 0.10 mole of atomic chlorine at equilibrium?

16. The equilibrium constant for the reaction $2 I(g) \rightleftarrows I_2(g)$ is 100 at about 1200°K. What volume vessel is required in order to have 25.4 g I_2 and 6.35 g I present at equilibrium?

17. All dissociation equilibria of the type $A_2(g) \rightleftarrows 2 A(g)$ are shifted to the right at higher temperatures. How does thermodynamics explain this fact?

9 RATES OF REACTION

Consider a general chemical reaction: $A + B \rightleftarrows AB$. What can we actually understand about the reaction? We can interpret our observations by atomic theory and label A, B, and AB as atoms, molecules, or ions in the gas, liquid, or solid phase. We can determine the structures and physical properties of the reactants and products. We can study the energy changes in the reaction, note where and if it reaches an equilibrium, and appreciate how entropy changes affect the reaction. What else is there? One thing we haven't mentioned is the rate of the reaction. We can learn a great deal by studying how fast A and B react to give AB.

Imagine $A + B \rightleftarrows AB$ representing some specific familiar reaction, such as the oxidation or iron:

$$2 \, Fe(s) + 3/2 \, O_2 \, (g) \rightleftarrows Fe_2O_3 \, (s)$$

If we look up some thermodynamic data for the reaction, we find that ΔH for the reaction is -196.5 kcal per mole, the entropy change is negative, and ΔG for the reaction is -177 kcal per mole (at 25°C). We conclude that the reaction is highly exothermic and, in spite of the negative entropy change, there is a large driving force toward spontaneous reaction. Yet, from experience we know that if we expose a piece of iron to the air it does not instantly rust. The key to understanding this apparent contradiction is to differentiate between the terms "spontaneous" and "instantaneous." Thermodynamics tells us that any reaction with a negative ΔG will occur spontaneously, but it reveals nothing about the length of time it will take for the reaction to occur. If finely divided iron is sprinkled into a flame, it ignites, and oxidation is instantaneous. ΔG for the reaction is approximately -177 kcal/mole and ΔH is -196.5 kcal/mole. If a piece of iron is exposed to the air, it rusts slowly over a period of years. ΔG for the reaction is still approximately -177 kcal/mole and ΔH is -196.5 kcal/mole. Thermodynamics does not differentiate between a fast reaction and a slow reaction. There are literally thousands of chemical reactions that are thermodynamically spontaneous, but are not instantaneous. So, what factors do influence the rate of a chemical reaction? What can we learn from reaction rates? The field of chemistry that seeks answers to these questions is called *chemical kinetics*.

For convenience, we shall divide the problems in this chapter into two categories.

Problem Categories

    I Factors that affect reaction rates.
    II Interpretation of reaction rates.

FACTORS THAT AFFECT REACTION RATES

Chemical kinetics is concerned with the rate of a chemical reaction. Reaction rates are influenced by (1) concentrations of reactants, (2) temperatures, (3) catalysts, and (4) the nature of the reactants.

For a general reaction $A + B \rightarrow C$ we can describe the rate at which the reaction proceeds in terms of the disappearance of A or B, or the formation of C. The rate of a reaction is proportional to the concentrations of the reactant species. As a mathematical equation we can state that

$$\text{rate} = \frac{-d(\text{conc. A})}{dt} \approx \frac{-\Delta(\text{conc. A})}{\Delta t} = k(\text{conc. A})^m (\text{conc. B})^n$$

or simply,

$$\text{rate} = k(\text{conc. A})^m (\text{conc. B})^n$$

This expression is called a *rate law*. The proportionality constant, k, is called the specific rate constant, and m and n are numerical exponents. The exponents m and n may be whole numbers, fractions, zero, or even negative numbers. The rate constant and the exponents m and n are parameters that must be determined by experiment.

Example 1.

A mixture of S_2Cl_2 (g) and Cl_2 (g) was initially 0.50 M in each gas inside a closed container. Find the initial rate of the reaction S_2Cl_2 (g) + Cl_2 (g) $\rightarrow$ S_2Cl_4 (g) from the following data:

Conc. Cl_2	Time
0.50 M	0
0.49 M	5 sec
0.48 M	10 sec

Solution.

By inspection, it is obvious that the rate of disappearance of Cl_2, and hence the rate of formation of S_2Cl_4, is 0.02 mole per liter every 10 seconds, or

$$\text{rate} = 0.02 \text{ M}/10 \text{ sec} = 0.002 \text{ M-sec}^{-1}$$

Example 2.

In separate experiments, the following data were obtained for the reaction in Example 1.

Experiment	conc. S_2Cl_2	conc. Cl_2	rate (M-sec^{-1})
1	0.50	0.50	0.002
2	0.50	0.25	0.001
3	0.50	1.00	0.004
4	0.25	0.50	0.0005
5	1.00	0.50	0.008

Determine the specific rate constant and the exponents in the rate law:

$$\text{rate} = k(\text{conc. } S_2Cl_2)^m (\text{conc. } Cl_2)^n$$

Solution.

Examine the first three sets of data. The conc. S_2Cl_2 is held constant and conc. Cl_2 is varied. When conc. S_2Cl_2 is constant, the entire term $k(\text{conc. } S_2Cl_2)^m$ is constant in the rate law, and the reaction rate depends only on $(\text{conc. } Cl_2)^n$. Note what happens as conc. Cl_2 is changed. Going from experiment 1 to experiment 2 the conc. Cl_2 is decreased by a factor of 2 and the rate decreases by a factor of 2. From experiment 1 to experiment 3, conc. Cl_2 is doubled and the rate doubles. The rate is directly proportional to conc. Cl_2 to the 1st power, i.e., the exponent n is 1.0. Another way of stating this is to say "the reaction is first order with respect to Cl_2."

Now examine the data from experiments 1, 4, and 5. In these experiments, conc. Cl_2 is held constant and conc. S_2Cl_2 is varied. When conc. S_2Cl_2 is decreased by a factor of 2, the rate decreases by a factor of 4. When conc. S_2Cl_2 is increased by 2, the rate goes up by 4. In other words, $(2)^m = 4$, and the exponent m must be 2.0. There is a squared dependence of the reaction rate on conc. S_2Cl_2, or "the reaction is second order with respect to S_2Cl_2."

Thus, the overall rate law must be

$$\text{rate} = k(\text{conc. } S_2Cl_2)^2 (\text{conc. } Cl_2)^1$$

The reaction is second order in S_2Cl_2, first order in Cl_2, and third order overall. The overall order is just the sum of the exponents.

To find the rate constant, k, we need only substitute the data from any experiment into the rate law. For example, using the data from experiment 1, we find

$$\text{rate} = k(\text{conc. } S_2Cl_2)^2 (\text{conc. } Cl_2)^1$$

$$(0.002 \text{ M-sec}^{-1}) = k(0.50)^2 (0.50)$$

$$k = \frac{0.002 \text{ M-sec}^{-1}}{0.125 \text{ M}^3} = 1.60 \times 10^{-2} \text{ M}^{-2}\text{-sec}^{-1}$$

Example 3.

For the reaction $Br_2 + Cl_2 \rightarrow 2\,BrCl$ the rate is found to be 0.050 M-sec^{-1} when conc. Br_2 is 0.20 M and conc. Cl_2 is 0.30 M. Find the rate constant, k, if the reaction were
    (a)   first order in Br_2 and first order in Cl_2.
    (b)   first order in Br_2 and second order in Cl_2.
    (c)   zero order in Br_2 and first order in Cl_2.

Solution.

(a)   rate = k(conc. Br_2) (conc. Cl_2) (exponent 1 understood)

$$k = 0.050/(0.20)\,(0.30) = 0.833 \text{ M}^{-1}\text{-sec}^{-1}$$

(b)   rate = k(conc. Br_2) (conc. Cl_2)^2

$$k = 0.050/(0.20)\,(0.30)^2 = 2.78 \text{ M}^{-2}\text{-sec}^{-1}$$

(c)   rate = k(conc. Br_2)^0 (conc. Cl_2)

$$k = 0.050/(1)\,(0.30) = 0.167 \text{ sec}^{-1}$$

Some chemical reactions and all radioactive decay processes obey a pure first order rate law, i.e., for A → products,

$$\text{rate} = k(\text{conc. A})$$

Many chemical reactions obey a pure second order rate law of the form

$$\text{rate} = k(\text{conc. A})^2$$

If we let X stand for conc. A, integration of these two rate laws yields the equation

$$\log \frac{X_0}{X} = \frac{kt}{2.30} \tag{1}$$

for a first order reaction, and

$$\frac{1}{X} - \frac{1}{X_0} = kt \tag{2}$$

for a second order reaction. The term X_0 is the concentration of A at some time zero, and X is the amount left after a time t. Equations (1) and (2) are useful in finding the amount of a substance remaining after some elapsed reaction time, and vice versa. Experimentally, rate laws are often determined from their integrated form.

Example 4.

Decay of the radioactive isotope $^{90}_{38}Sr$ obeys a first order rate law. The rate constant, k, is 3.48×10^{-2} years^{-1}. Starting with 1.0 gram of $^{90}_{38}Sr$, find
- (a)   the "half-life," or the time required for ½ of the $^{90}_{38}Sr$ to undergo radioactive decay.
- (b)   the amount of $^{90}_{38}Sr$ remaining after 30 years.

Solution.

(a)   Equation (1) was derived using concentration of A; however, since X_0/X is a ratio, it is possible to use moles, grams, etc. directly in the equation. For this problem we can set the initial amount of A equal to 1.00 and the amount remaining after one "half-life" equal to 0.50.

$$\log \frac{X_0}{X} = \frac{kt}{2.30}; \log \frac{(1.00)}{(0.50)} = \frac{(3.48 \times 10^{-2} \text{ yrs}^{-1}) \times t_{1/2}}{2.30}$$

$$t_{1/2} = \frac{(0.301)(2.30)}{3.48 \times 10^{-2} \text{ yrs}^{-1}} = 19.9 \text{ years}$$

Note that for any first order reaction, $\log 1.00/0.50$ is a constant, hence the "half-life" is independent of initial amount and depends only on k. The relationship is

$$t_{1/2} = \frac{0.693}{k}$$

(b) We can substitute into equation (1) using 1.00 gram as initial amount and X for the amount remaining after 30 years.

$$\log \frac{1.00 \text{ g}}{X} = \frac{(3.48 \times 10^{-2} \text{ yrs}^{-1})(30 \text{ yrs})}{2.30},$$

$$\log \frac{1.00}{X} = 0.454; \frac{1.00}{X} = 2.84$$

$$X = 0.35 \text{ grams remaining after 30 yrs}$$

Example 5.

The decomposition of $NOCl(g)$ ($2 \, NOCl(g) \rightleftarrows 2 \, NO(g) + Cl_2(g)$) is pure second order with a rate constant of $9.01 \times 10^{-4} \, M^{-1}\text{-min}^{-1}$ at $300°K$. Beginning with conc. $NOCl = 0.40 \, M$, find
(a) the half-life for the decomposition.
(b) the amount of Cl_2 formed after 5.0 hours.

Solution.

(a) After ½ of the NOCl has reacted, conc. NOCl will be 0.20 M, and we can substitute into equation (2):

$$\frac{1}{X} - \frac{1}{X_0} = kt; \frac{1}{0.20 \text{ M}} - \frac{1}{0.40 \text{ M}} = (9.01 \times 10^{-4} \, M^{-1}\text{-min}^{-1}) \times t_{1/2};$$

$$t_{1/2} = \frac{1}{(0.40 \text{ M})(9.01 \times 10^{-4} \, M^{-1}\text{-min}^{-1})} = 2.77 \times 10^3 \text{ minutes}$$

Can you show that, for any pure second order reaction, $t_{1/2} = 1/X_0 k$?
(b) We can substitute into equation (2) after converting 5.0 hrs to 300 minutes.

$$\frac{1}{X} - \frac{1}{X_0} = kt; \frac{1}{X} - \frac{1}{0.40 \text{ M}} = (9.01 \times 10^{-4} \, M^{-1}\text{-min}^{-1}) \times 300 \text{ min}.$$

$$\frac{1}{X} = 0.27 \, M^{-1} + 2.50 \, M^{-1}; \quad X = 0.36 \, M$$

Note that X is the molar conc. NOCl remaining after 5.0 hrs. If 0.04 M of the NOCl have reacted, the amount of Cl_2 formed would be 0.02 mole per liter from the reaction

$$2 \, NOCl \rightleftarrows 2 \, NO + Cl_2$$

Having examined the effect of concentration on reaction rate, let's briefly consider the other factors that influence the rate of a reaction. There is not much to be said about how the nature of reactants affects the rate of reaction. Obviously different substances react at different rates. The reasons why are inherent in the fact that molecules, atoms, or ions interact chemically with each other by different mechanisms. It would be quite surprising if the rate of a reaction such as $Br_2 + Cl_2 \rightarrow 2\ BrCl$ were exactly the same as the rate of $2\ Fe + 3/2\ O_2 \rightarrow Fe_2O_3$.

Substances called *catalysts* can influence rates of chemical reactions. A catalyst is a material that speeds up (or slows down) a reaction without any apparent change occurring in the catalyst itself. Catalysts can operate in various ways, but apparently their principal function is to afford an alternate pathway by which a chemical reaction can proceed (see Category II, Example 3).

An increase in temperature always increases the rate of a chemical reaction. A 10°C increase in temperature will approximately double the rate of a reaction. The theoretically derived and experimentally tested equation that relates rate of reaction to temperature is

$$\log \frac{k_2}{k_1} = \frac{Ea}{2.30\ R} \left(\frac{T_2 - T_1}{T_2 T_1} \right)$$

where k_1 and k_2 are rate constants at absolute temperatures T_1 and T_2, and Ea is a parameter called the *activation energy*. The equation is based on gas phase studies, and R is the gas law constant. If R is taken at 1.99 cal/°K, the equation becomes

$$\log \frac{k_2}{k_1} = \frac{Ea}{4.58\ cal/°K} \left(\frac{T_1 - T_1}{T_2 T_1} \right)$$

The activation energy, Ea, is an energy "barrier" that atoms, ions, or molecules must cross before they react (see Category II).

Example 6.

For the second order reaction $2\ NO_2 \rightarrow 2\ NO + O_2$, the activation energy is known to be 26.6 kcal per mole. The rate constant, k, is 2.94 $M^{-1} sec^{-1}$ at 900°K. (a) Find the rate constant at 300°K. (b) What would the activation energy have to be if the reaction rate exactly doubled from 300°K to 310°K?

Solution.

(a) $\log \dfrac{k_2}{k_1} = \dfrac{Ea}{4.58\ cal/°K} \left(\dfrac{T_2 - T_1}{T_2 T_1} \right) = \dfrac{26,600\ cal/mole}{4.58\ cal/°K} \left(\dfrac{900°K - 300°K}{900°K \times 300°K} \right),$
setting $T_2 = 900°K$ and $T_1 = 300°K$

$\log \dfrac{2.94}{k_1} = (5805)\ (2.22 \times 10^{-3})$; $\log 2.94 - \log k_1 = 12.907$

$\log k_1 = -12.439$; $k_1 = 3.64 \times 10^{-13}\ M^{-1}\text{-}sec^{-1}$

(b) $\log \dfrac{2 \times 3.64 \times 10^{-13}}{3.64 \times 10^{-13}} = \dfrac{Ea}{4.58} \left(\dfrac{310 - 300}{310 \times 300} \right)$; $Ea = 12,822$

CATEGORY II · INTERPRETATION OF REACTION RATES

According to collision theory, atoms, ions, or molecules must collide with each other or the walls of a container before existing chemical bonds can be broken and/or new bonds formed. It is collision that leads to chemical reaction, but not all collisions result in reaction. The colliding species must have the proper orientation and some minimum kinetic energy (the activation energy) before reaction can occur. It has been estimated that only about one in every 10^{14} collisions actually results in reaction. The rate of a reaction must, therefore, depend upon the total number of collisions taking place in a given time and upon the magnitude of the activation energy. The greater the activation energy for a reaction, the slower the reaction will proceed. Increasing the concentrations of reactants will increase the number of collisions between the reacting species, and therefore increase the rate of reaction. Increasing the temperature will increase the number of molecules that have sufficient kinetic energy to equal or exceed the activation energy, and adding a catalyst lowers the activation energy by providing some easier pathway by which the reaction can proceed. If the premises of collision theory are correct, then studies of the rate of a reaction should allow us to make reasonable guesses about the *mechanism* of the reaction.

The mechanism of a reaction is simply a molecular picture of how the reaction takes place. For example, if we start with H_2 and O_2 and end up with H_2O, what actually happens to the molecules in the course of the reaction? The answer to that question would give us a mechanism for the reaction. A reaction may proceed in one, or two, or several steps. Whatever the case, the overall rate of the reaction must be limited by the slowest step in the mechanism. The form of the rate law indicates the composition of the intermediate formed in the slowest step in the mechanism. That intermediate species is often called the activated complex. Knowing the composition of the intermediate formed in the slowest step in the reaction enables us to make an educated guess concerning the reaction mechanism.

Example 1.

For the reaction $NO(g) + O_3(g) \rightarrow NO_2(g) + O_2(g)$ the rate law is rate = k (conc. NO) (conc. O_3). (a) What is the slowest step in the reaction? (b) Propose a plausible mechanism for the reaction.

Solution.

(a) The exponents in the rate law tell us the number of each species making up the intermediate in the slowest step. In this case, the reaction is 1st order in NO and 1st order in O_3; therefore, the slowest step in the reaction must be

$$NO + O_3 \rightarrow (NO \cdot O_3)^*$$

We can tell nothing for certain about the short-lived, activated (high-energy) species $(NO \cdot O_3)^*$ except its composition. Knowing the structures of NO_2 and O_3, it is *probable* that the intermediate has a structure such as

$$O \cdots O \cdots O \cdots N \cdots O$$

Then, rupture at the arrow would result in formation of NO_2 and O_2 molecules.

(b) Once the slowest step is accounted for, it becomes anybody's guess as to the rest of the mechanism, so long as the sum of the steps add up to give the balanced equation for the reaction. Also, common sense would tend to rule out some really "wild" mechanism — for example, simultaneous 3-, or 4-, or 5-body collisions would be much less probably than two-body collisions. In this example, there is no reason to suppose that the mechanism involves anything other than the collision of NO with O_3 to give $(NO \circ O_3)^*$ which then decomposes to form NO_2 and O_2:

$$NO + O_3 \rightarrow (NO \cdot O_3)^*$$

$$(NO \circ O_3)^* \rightarrow NO_2 + O_2$$

Example 2.

For the reaction $2\,NO(g) + 2\,H_2(g) \rightarrow N_2(g) + 2\,H_2O(g)$ it is known that H_2O_2 is an intermediate. The rate law is

$$rate = k\,(conc.\ NO)^2\,(conc.\ H_2)$$

Is the following mechanism consistent with the rate law?

$$NO + NO \rightarrow (NO)_2 \quad fast$$

$$(NO)_2 + H_2 \rightarrow (2\,NO \cdot H_2)^* \quad slow$$

$$(2\,NO \cdot H_2)^* \rightarrow N_2 + H_2O_2$$

$$H_2O_2 + H_2 \rightarrow 2\,H_2O \quad fast$$

Solution.

The second (slow) step does involve formation of an intermediate from 2 NO and one H_2. Therefore, the mechanism *is* consistent with the rate law. The different steps do add to give the balanced equation, so while we cannot be certain that the mechanism is correct, we cannot say that it is wrong.

Example 3.

The oxidation of $SO_2(g)$ to give $SO_3(g)$ proceeds slowly at room temperature. A trace of $NO(g)$ acts as a catalyst and speeds up the reaction. In the presence of NO, the reaction obeys the rate law:

$$rate = k(conc.\ NO)^2\,(O_2)$$

Propose a mechanism for the catalyzed reaction.

Solution.

The balanced equation is

$$2\,SO_2 + O_2 \rightarrow 2\,SO_3$$

In order to be called a catalyst, the NO must somehow "get into the act," yet be recovered unchanged after the reaction. A plausible mechanism might be

$$NO + NO \rightarrow (NO)_2 \quad fast$$

$$(NO)_2 + O_2 \rightarrow (2\,NO \cdot O_2)^*$$

$$(2\,NO \cdot O_2)^* \rightarrow 2\,NO_2$$

$$2\,NO_2 + 2\,SO_2 \rightarrow 2\,SO_3 + 2\,NO \quad (fast)$$

Note: Adding the steps gives the desired balanced equation and shows that the NO would be recovered "unchanged." By providing an easier (hence faster) pathway for the reaction, the NO has acted as a catalyst. Can you suggest a possible structure for the intermediate $(2\,NO \cdot O_2)^*$?

1. In solution, ammonium cyanate undergoes a spontaneous rearrangement to give urea according to the reaction $NH_4CNO \rightarrow (NH_2)_2CO$. An experiment is conducted in which the initial concentration of NH_4CNO is 0.40 M. The concentration of urea is determined at different times. The results are

Time (min)	Conc. $(NH_2)_2CO$
10	0.076 M
25	0.147 M
40	0.193 M
60	0.233 M

Find the rate law and the specific rate constant for the reaction.

2. For the reaction

$$H_3PO_4 + 3I^- + 2H^+ \rightleftarrows H_3PO_3 + I_3^- + H_2O$$

initial rates of reaction were determined for different initial concentrations. The results are listed below. Determine the rate law for the reaction.

Rate (M-sec^{-1})	Conc. H_3PO_4	Conc. I$^-$	Conc. H$^+$
2.8×10^{-7}	0.01	0.20	0.10
5.6×10^{-7}	0.01	0.40	0.10
5.6×10^{-7}	0.02	0.20	0.10
2.8×10^{-7}	0.02	0.40	0.05
2.24×10^{-6}	0.02	0.20	0.20

3. At a certain temperature, the half-life for the decomposition of xenon difluoride $(XeF_2(g) \rightarrow Xe^{\circ}(g) + F_2(g))$ is 200 seconds when the initial concentration of XeF_2 is 0.01 M, and 66.7 seconds when the initial concentration is 0.03 M. What is the order of the reaction and the specific rate constant?

4. When the initial concentration of thiocyanate ion, SCN^-, is 0.010 M in aqueous solution at 70°C, the half-life for the decomposition reaction $SCN^- \rightarrow S(s) + CN^-$ is 26.0 min. How long will it take for conc. SCN^- to drop to 0.002 M if the reaction is
  (a) first order (b) second order

5. The first order decomposition of N_2O_5 to give NO_2 and O_2 has a specific rate constant of 4.30×10^{-5} sec^{-1} at 300°K and 6.97×10^2 sec^{-1} at 500°K. Calculate the activation energy for this reaction.

6. For the reaction $A + B \rightarrow$ products, the initial rates of reaction at various concentrations of A and B are

rate (M-sec^{-1})	conc. A	conc. B
1.5×10^{-2}	0.20	0.20
1.5×10^{-2}	0.20	0.40
6.0×10^{-2}	0.40	0.40

Find the order with respect to A and B.

7. Radioactive decay is a first order process. The half-life of the radioactive isotope $^{235}_{92}U$ is 7.13×10^8 years. If a rock contained 1.00 g of $^{235}_{92}U$ when the earth was formed and the age of the earth is 4.5 billion years, how much $^{235}_{92}U$ would be found in the rock today?

8. For a reaction A + B → products, the activation energy is known to be 12,000 cal/mole. At 25.0°C the specific rate constant is 1.20×10^{-1} M^{-1}-sec^{-1}. At what temperature would the rate constant become equal to 9.6×10^{-1} M^{-1}-sec^{-1}?

Category II

9. What is the slowest step in the mechanism of the reaction in Problem 2? Propose a reasonable mechanism for the reaction.

10. For the reaction $A + B_2 \xrightarrow{C} AB + B$, the rate law is rate = k(conc. B_2) (conc. C), where C is a catalyst. Propose a mechanism consistent with these observations.

11. For the reaction, A + B → C, find the rate law if the mechanism is

$$2\,B \to B_2 \text{ (fast)}$$

$$A + B_2 \to (A \cdot B_2)^* \text{ (slow)}$$

$$(A \circ B_2)^* \to B + C \text{ (fast)}$$

12. For the substitution reaction

$$[Co(en)_2Cl_2]^+ + Br^- \to [Co(en)_2BrCl]^+ + Cl^-$$

the reaction is found to be first order in $[Co(en)_2Cl_2]^+$ and zero order in Br^-. Propose a plausible mechanism for the reaction.

General Problems

13. At a certain temperature the rate constant for the reaction $2\,ICl + H_2 \to I_2 + 2\,HCl$ is 1.62×10^{-1} liter/mole-sec. The rate law is

$$\text{rate} = k(\text{conc. } ICl)\,(\text{conc. } H_2)$$

What is the rate of the reaction at the following concentrations?

Molarity ICl	Molarity H_2
0.10	0.10
0.20	0.10
0.20	0.40

14. A compound decomposes by a first order reaction. After 4 hrs only 25% of the original compound remains. Find the specific rate constant, k, and the half-life for the reaction.

15. For the reaction 2 B → C, the following data were obtained:

Molar conc. B	1.00	0.862	0.758	0.678	0.609
Time (min)	0	500	1000	1500	2000

Determine whether the reaction is first order or second order in B and find the specific rate constant.

16. A reaction has an activation energy of 25 kcal and a specific rate constant $k = 5.0 \times 10^{-2}$ sec^{-1} at 400°K. Find the specific rate constant at 500°K.

17. For the reversible reaction A $\rightleftarrows$ B, the forward reaction is first order in A with a specific rate constant k_f. The reverse reaction is first order in B with a specific rate constant k_r. Derive a relationship between k_f, k_r, and K, the equilibrium constant for the reaction.

10 SOLUBILITY EQUILIBRIA

Although no solute is ever 100% insoluble in a particular solvent, many substances appear virtually insoluble or are very slightly soluble. How much one substance dissolves in another depends upon the polarity of the solute and solvent, the nature of the bonding in the solute, temperature, and several other factors. A chemist frequently needs to know exactly how much of a compound will dissolve in a given solvent and what factors affect the solubility. Solubility equilibria of slightly soluble ionic compound have been extensively studied.

Consider a saturated aqueous solution of an ionic compound. The solution is saturated when as much of the compound dissolves in the water as is possible at a given temperature. In fact, imagine that there is some of the solid compound, undissolved, in the bottom of the container. In the saturated solution there is an equilibrium established between the dissolved ions in solution and the undissolved solid. If we represent the ionic compound by the general formula A^+B^-, where there are equal numbers of cations and anions, the equilibrium may be written as

$$A^+B^-(\text{solid}) \rightleftarrows A^+(\text{aqueous}) + B^-(\text{aqueous})$$

As with any equilibrium system, this is a "two way street" with ions going into solution at the same rate that aqueous ions recombine and precipitate as undissolved solid. Also, as is true of any equilibrium, we can write an equilibrium constant expression:

$$\frac{[A^+(\text{aq})]\ [B^-(\text{aq})]}{[A^+B^-(\text{s})]} = K$$

Characteristic of this type of equilibrium is the appearance of $[A^+B^-(\text{s})]$ in the denominator of the equilibrium constant equation. Since the [moles per liter] of a *solid* is in itself a constant quantity, we can simplify the expression by multiplying both sides of the equation by $[A^+B^-(\text{s})]$; thus,

$$[A^+(\text{aq})]\ [B^-(\text{aq})] = K \times [A^+B^-(\text{s})] = K_{sp},$$

where K_{sp} is given the special name "solubility product constant." If the compounds were AgOH, CdS, or AgI, for example, the solubility product constants would be

$$[Ag^+]\ [OH^-] = K_{sp}$$

$$[Cd^{++}]\ [S^=] = K_{sp}$$

$$[Ag^+]\ [I^-] = K_{sp}$$

in which the subscript (aqueous) or (aq) is eliminated for brevity.

Similarly, when the ratios of positive and negative ions are not one to one, we have the general solubility equilibrium

$$A_mB_n(s) \rightleftarrows mA^{+n}(aq) + nB^{-m}(aq)$$

and the K_{sp} expression (equation) is

$$[A^{+n}]^m [B^{-m}]^n = K_{sp}$$

For example, the solubility product constants for CaF_2, Fe_2S_3, and Ag_2S would be

$$[Ca^{+2}] [F^-]^2 = K_{sp}$$

$$[Fe^{+3}]^2 [S^{-2}]^3 = K_{sp}$$

$$[Ag^+]^2 [S^{-2}] = K_{sp}$$

Table 10–1 lists some solubility product constants.

Problem Categories

The chemist is usually faced with three types of problems involving solubility equilibria of slightly soluble ionic compounds.

I   What are the maximum ionic concentrations in solutions of "pure," slightly soluble compounds?

II How much of one ion can exist with a known amount of another ion in solution?

Table 10–1   Solubility Product Constants (25°C)

Acetates			**Fluorides**	
$AgC_2H_3O_2$	2.0×10^{-3}		CaF_2	2.0×10^{-10}
			PbF_2	4.0×10^{-8}
Bromides			BaF_2	1.7×10^{-6}
$AgBr$	5.0×10^{-13}			
$CuBr$	4.9×10^{-9}		**Hydroxides**	
$PbBr_2$	5.0×10^{-6}		$Fe(OH)_3$	5.0×10^{-38}
			$Al(OH)_3$	5.0×10^{-33}
Carbonates			$Fe(OH)_2$	1.0×10^{-15}
Ag_2CO_3	6.2×10^{-12}		$Mg(OH)_2$	3.2×10^{-11}
$BaCO_3$	2.0×10^{-9}		$Ca(OH)_2$	5.5×10^{-5}
$CaCO_3$	8.7×10^{-9}			
$ZnCO_3$	3.0×10^{-8}		**Iodides**	
Li_2CO_3	1.7×10^{-5}		AgI	1.0×10^{-16}
			PbI_2	2.4×10^{-8}
Chlorides				
$AgCl$	1.6×10^{-10}		**Sulfates**	
$PbCl_2$	2.4×10^{-4}		$BaSO_4$	1.1×10^{-10}
			Ag_2SO_4	1.6×10^{-5}
Chromates			$CaSO_4$	6.4×10^{-5}
Ag_2CrO_4	2.2×10^{-12}			
$BaCrO_4$	3.0×10^{-10}		**Sulfides**	
			HgS	1.0×10^{-50}
			Ag_2S	1.0×10^{-49}
			Cu_2S	1.0×10^{-46}
			PbS	4.2×10^{-28}
			MnS	1.4×10^{-15}

III When will a substance stay in solution, and when will it precipitate as a solid?

ION CONCENTRATIONS OF PURE, SLIGHTLY SOLUBLE COMPOUNDS   CATEGORY

Example 1.

The K_{sp} of silver iodide, AgI, is 1.0×10^{-16}. What is the $[Ag^+]$ in a saturated solution of AgI?

Solution.

For all equilibrium problems it is a good idea to write down first, the specific equilibrium being considered, and second, the equation for the equilibrium constant expression. In this case the equilibrium is

$$AgI(s) \rightleftarrows Ag^+ + I^-$$

and the equilibrium constant is the K_{sp} equation:

$$[Ag^+]\ [I^-] = 1.0 \times 10^{-16}$$

Now, since we have only pure AgI to consider, we see that as it goes into solution every time there is a silver ion, Ag^+, produced, there must also be an iodide ion, I^-, released at the same time. At equilibrium in the saturated solution, the number of Ag^+ ions must be the same as the number of I^- ions. In terms of concentration, $[Ag^+] = [I^-]$. We know that $[Ag^+] \times [I^-]$ will be 1.0×10^{-16}, so we can apply a little simple algebra, with $[Ag^+]$ being the unknown, X. From the equilibrium, we know that if $[Ag^+] = X$, $[I^-]$ is also X.

$$\overset{X}{} \quad \overset{X}{}$$
$$AgI(s) \rightleftarrows Ag^+ + I^-$$

Then, from the K_{sp} expression, we have an equation with only one unknown, X.

$$\overset{X}{} \quad \overset{X}{}$$
$$[Ag^+]\ [I^-] = 1.0 \times 10^{-16}$$

$$X^2 = 1.0 \times 10^{-16}$$

$$X = 1.0 \times 10^{-8}$$

So, the problem is solved. The $[Ag^+]$ in a saturated AgI solution is 1.0×10^{-8} M. What is the iodide ion concentration? It is exactly the same, isn't it?

Example 2.

If the maximum amount of zinc carbonate, $ZnCO_3$, that will dissolve to give 1.00 liter of solution is 0.0213 gram, what is the K_{sp} of $ZnCO_3$?

Solution.

Here, the equilibrium is

$$ZnCO_3\ (s) \rightleftarrows Zn^{2+} + CO_3^{2-}$$

and the solubility product constant is

$$[Zn^{2+}]\ [CO_3^{2-}] = K_{sp}$$

To calculate the K_{sp}, we need to know $[Zn^{2+}]$ and $[CO_3^{2-}]$ in the saturated solution. Once again, since we have only the pure compound $ZnCO_3$ for every $ZnCO_3$ (s) going into solution, we get a Zn^{2+} ion and a CO_3^{2-} ion and at equilibrium, $[Zn^{2+}] = [CO_3^{2-}]$. The grams (per liter) of $ZnCO_3$ that dissolve = .0213, so the moles (per liter) = grams ÷ formula wt = .0213 ÷ 125.4 = 1.70×10^{-4}. This is the number of moles per liter of $ZnCO_3$ that dissolves, or the *molar solubility* of zinc carbonate. Thus, in the solution, $[Zn^{2+}] = [CO_3^{2-}] = 1.70 \times 10^{-4}$ M, and the K_{sp} is

$$(1.70 \times 10^{-4})(1.70 \times 10^{-4}) = 2.89 \times 10^{-8}$$

Example 3.

Calcium fluoride, CaF_2, is used in the "fluoridation" of water. If the molar solubility of CaF_2 is 2.0×10^{-4} M, calculate the K_{sp} of CaF_2.

Solution.

The equilibrium is

$$CaF_2 \,(s) \rightleftarrows Ca^{2+} + 2\,F^-$$

and the K_{sp} expression is

$$[Ca^{2+}]\,[F^-]^2 = K_{sp}$$

The important thing to note here is that as CaF_2 (s) dissolves, there are *two* F^- ions produced for every one Ca^{2+} ion! If 2.0×10^{-4} mole per liter of CaF_2 (s) dissolves, this produces 2.0×10^{-4} mole per liter of Ca^{2+} ions and twice that amount, or 4.0×10^{-4} mole per liter of F^- ion. Substituting into the K_{sp} expression we find

$$(2.0 \times 10^{-4})(4.0 \times 10^{-4})^2 = 3.2 \times 10^{-11}$$

Example 4.

Given that the K_{sp} of CaF_2 is 3.2×10^{-11}, calculate $[Ca^{2+}]$ in a saturated CaF_2 solution.

Solution.

Keep in mind that for pure CaF_2 there will be two F^- ions for each Ca^{2+} ion. If we define $[Ca^{2+}]$ as the unknown, call it "X," then $[F^-]$ will be twice that or "2X."

$$\overset{\displaystyle X \quad\quad 2X}{CaF_2 \,(s) \rightleftarrows Ca^{2+} + 2\,F^-}$$

Substituting into the K_{sp} equation, we have $[Ca^{2+}] = X$; $[F^-] = 2X$, so

$$(X)(2X)^2 = 3.2 \times 10^{-11}$$

$$4X^3 = 3.2 \times 10^{-11}$$

$$X^3 = 0.8 \times 10^{-11} = 8.0 \times 10^{-12}$$

$$X = \sqrt[3]{8.0 \times 10^{-12}} = 2.0 \times 10^{-4}$$

This is the $[Ca^{2+}]$. What is $[F^-]$?

CONCENTRATION OF ONE ION WHEN THAT OF ANOTHER IS KNOWN　CATEGORY

Example 1.

The K_{sp} of $BaSO_4$ is 1.10×10^{-10}. What is the maximum $[Ba^{2+}]$ that can exist in a 0.50 M Na_2SO_4 solution?

Solution.

We know that

$$[Ba^{2+}] \ [SO_4{}^{2-}] = 1.10 \times 10^{-10}$$

The K_{sp} expression tells us that in a saturated solution of pure $BaSO_4$, $[Ba^{2+}]$ times $[SO_4{}^{2-}]$ is 1.10×10^{-10}. It also tells us that in any solution containing Ba^{2+} and $SO_4{}^{2-}$ the product of $[Ba^{2+}]$ times $[SO_4{}^{2-}]$ cannot be greater than 1.10×10^{-10}. If we tried to make it greater, what would happen? If we tried to push the "ion product" $[Ba^{2+}] \times [SO_4{}^{2-}]$ beyond the K_{sp} limit, solid $BaSO_4$ would precipitate out! Remember, that "[ ]" expression refers to the concentration of an ion *in the solution*. So, K_{sp} sets a limit on how much of one ion can exist *in solution* in the presence of another ion. In this case the $[SO_4{}^{2-}]$ in the solution is 0.50 M from the Na_2SO_4. Therefore, the maximum that the $[Ba^{2+}]$ could be is

$$[Ba^{2+}] = \frac{1.10 \times 10^{-10}}{[SO_4{}^{2-}]} = \frac{1.10 \times 10^{-10}}{0.50} = 2.20 \times 10^{-10}$$

Could $[Ba^{2+}]$ in solution be less than this?

Example 2.

The K_{sp} of MgF_2 is 6.4×10^{-9}. What is the maximum concentration of F^- ion possible in a solution that is 0.01 M in Mg^{2+}?

Solution.

The equilibrium is

$$MgF_2 \ (s) \rightleftarrows Mg^{2+} + 2 \ F^-$$

and the K_{sp} expression is

$$[Mg^{2+}] \ [F^-]^2 = 6.4 \times 10^{-9}$$

The problem here is similar to Example 1, but another way of looking at it would be by applying Le Chatelier's Principle or what is sometimes called the "common ion" effect. For this system, at equilibrium, what would happen to the number of Mg^{2+} ions in solution if we increased the number of F^- ions? The equilibrium would "shift" to the left and the number of Mg^{2+} ions would have to decrease (or be taken out of solution as solid MgF_2 precipitated). Conversely, increasing the concentration of Mg^{2+} would have to decrease the concentration of F^-. So the amount of one ion in the solution dictates the amount (concentration) of the other ion which can co-exist with it. In this problem the $[Mg^{2+}]$ is 0.01 M, so we can find the maximum $[F^-]$ by substituting into the K_{sp} equation:

$$[F^-]^2 = \frac{6.4 \times 10^{-9}}{[Mg^{2+}]} = \frac{6.4 \times 10^{-9}}{(0.01)} = 6.4 \times 10^{-7}$$

or, for convenience, $[F^-]^2 = 64 \times 10^{-8}$
and, $[F^-] \ \ = 8.0 \times 10^{-4}$

We find that in a solution 0.01 M in Mg^{2+} ion the F^- ion concentration in the solution could be as high as 8.0×10^{-4} M without precipitating MgF_2. It could be less than 8.0×10^{-4} M without exceeding the K_{sp} limit, but it could not be greater.

Notice the difference between this problem and Examples 3 and 4 of the Category I problems. If the Mg^{2+} ions and F^- ions in the solution were coming from just "pure" MgF_2, then $[F^-]$ would have to be exactly twice the $[Mg^{2+}]$. In this type of problem the Mg^{2+} and F^- are presumed to be from two different sources, hence the concentration of F^- need not be twice the Mg^{2+} concentration. Indeed, $[Mg^{2+}]$ and $[F^-]$ in the solution may have a wide range of values so long as the K_{sp} limitation is obeyed.

Example 3.

The K_{sp} of silver chloride, AgCl, is 1.0×10^{-10}. How many grams of NaCl could be added to 500 ml of a 0.20 M $AgNO_3$ solution without precipitating AgCl?

Solution.

Since $[Ag^+]$ is 0.20 M (from the $AgNO_3$), the maximum concentration of Cl^- without precipitation is

$$[Cl^-] = \frac{1.0 \times 10^{-10}}{[Ag^+]} = \frac{1.0 \times 10^{-10}}{(0.20)} = 5.0 \times 10^{-10} \text{ M}$$

If there could be 5.0×10^{-10} mole of Cl^- per liter, then in 500 ml there could only be 2.5×10^{-10} mole of Cl^-. Since this will come from NaCl and assuming no volume change when the NaCl is added, the number of grams of NaCl could be obtained by

$$Xg \text{ NaCl} = 2.5 \times 10^{-10} \text{ mole NaCl} \times \frac{58.5 \text{ g NaCl}}{1 \text{ mole NaCl}} = 1.46 \times 10^{-8} \text{ gram}$$

It is obvious why precipitation of AgCl by addition of $AgNO_3$ to a solution is a sensitive test for chloride ion!

CATEGORY III   DETERMINING WHEN AND IF A SUBSTANCE WILL PRECIPITATE

Example 1.

If 500 ml of 0.0001 M $AgNO_3$ were mixed with 500 ml of 0.00001 M NaCl solution, would AgCl precipitate? $K_{sp} = 1.0 \times 10^{-10}$.

Solution.

The limitation imposed by the solubility product constant tells us that in a solution containing Ag^+ and Cl^-, the $[Ag^+]$ times $[Cl^-]$ *in solution* can be as great as, but not greater than, 1.0×10^{-10}. If the "ion product" $[Ag^+] \times [Cl^-]$ is less than K_{sp}, then AgCl will not precipitate. If the ion product exceeds K_{sp}, precipitation will occur. In the final solution, after mixing, the $[Ag^+]$ from the $AgNO_3$ could be determined as follows:

$$M = \frac{\text{no. of moles}}{\text{no of liters}}; \text{ no. of moles} = M \times \text{no. of liters}$$

$$\text{no. of moles } Ag^+ = 0.0001 \text{ M} \times 0.500 \text{ liters}$$

$$= 5.0 \times 10^{-5} \text{ mole } Ag^+$$

$$[Ag^+] = \frac{\text{no. of moles } Ag^+}{\text{final no. of liters}} = \frac{5.0 \times 10^{-5} \text{ mole}}{(1.0 \text{ liter})} = 5.0 \times 10^{-5} \text{ M}$$

Similarly, the Cl^- ion from NaCl would be

$$\text{no. of moles } Cl^- = 0.00001 \text{ M} \times 0.50 \text{ liter}$$
$$= 5.0 \times 10^{-6} \text{ mole}$$

and,

$$[Cl^-] = \frac{5.0 \times 10^{-6} \text{ mole}}{(1.0 \text{ liter})} = 5.0 \times 10^{-6} \text{ M}$$

The "ion product" in the solution would be $[Ag^+] \times [Cl^-]$ which is equal to

$$(5.0 \times 10^{-5}) \times (5.0 \times 10^{-6}) = 2.5 \times 10^{-10}$$

and since the K_{sp} of AgCl is 1.0×10^{-10}, the ion product is slightly larger than the K_{sp} and some solid AgCl would precipitate. If the $[Ag^+]$ had been 5.0×10^{-6} M, for example, then the ion product would have been 2.5×10^{-11} and there would be no precipitate of AgCl.

Example 2.

The K_{sp} of $ZnCO_3$ is 3.0×10^{-8}. Will precipitation occur if 25.0 ml of 0.01 M $ZnCl_2$ solution is mixed with 125.0 ml of 0.003 M Na_2CO_3 solution?

Solution.

Once again, we must determine whether or not the ion product exceeds the K_{sp}. The number of moles of Zn^{2+} ion, from the $ZnCl_2$, will be

$$\text{no. of moles } Zn^{2+} = 0.01 \text{ M} \times 0.025 \text{ } \ell = 2.5 \times 10^{-4}$$

The number of moles of CO_3^{2-} ion, from Na_2CO_3, is

$$\text{no. of moles } CO_3^{2-} = 0.003 \text{ M} \times 0.125 \text{ } \ell = 3.75 \times 10^{-4}$$

Assuming the volumes of the solutions are additive, the final volume of the solution is 150 ml or 0.150 liter, and

$$[Zn^{2+}] = \frac{2.5 \times 10^{-4} \text{ mole}}{0.150 \text{ liter}} = 1.67 \times 10^{-3} \text{ M}$$

$$[CO_3^{2-}] = \frac{3.75 \times 10^{-4} \text{ mole}}{0.150 \text{ liter}} = 2.50 \times 10^{-3} \text{ M}$$

The ion product, $(1.67 \times 10^{-3}) \times (2.50 \times 10^{-3}) = 4.18 \times 10^{-6}$, does exceed the K_{sp} and $ZnCO_3$ would precipitate.

Example 3.

If 1.35 liters of water were added to the solution in Example 2, would the $ZnCO_3$ dissolve?

Solution.

Total volume of the solution would then be 1.35 liters plus 0.150 liter, or 1.50 liters. Thus, final concentrations would be

$$[Zn^{2+}] = \frac{2.5 \times 10^{-4} \text{ mole}}{1.5 \text{ liters}} = 1.67 \times 10^{-4} \text{ M}$$

$$[CO_3{}^{2-}] = \frac{3.75 \times 10^{-4} \text{ mole}}{1.5 \text{ liters}} = 2.50 \times 10^{-4} \text{ M}$$

and the ion product would be

$$(1.67 \times 10^{-4}) \times (2.50 \times 10^{-4}) = 4.18 \times 10^{-8}$$

The ion product still exceeds the K_{sp} and the conclusion is that all of the $ZnCO_3$ would not dissolve.

Example 4.

The K_{sp} of $PbCl_2$ is 2.4×10^{-4} and the K_{sp} of PbI_2 is 2.4×10^{-8}. If Pb^{2+} were added to a solution 0.01 M in Cl^- and 0.01 M in I^-, at what $[Pb^{2+}]$ would PbI_2 and $PbCl_2$ just begin to precipitate? What would the iodide ion concentration be at the point where $PbCl_2$ starts to precipitate?

Solution.

The point at which the solution will be saturated with respect to PbI_2 will be when the ion product, $[Pb^{2+}] \, [I^-]^2$, is just equal to the K_{sp} of PbI_2. For an $[I^-]$ of 0.01 M, this will occur when

$$[Pb^{2+}] = \frac{2.4 \times 10^{-8}}{[I^-]^2} = \frac{2.4 \times 10^{-8}}{(0.01)^2} = 2.4 \times 10^{-4} \text{ M}$$

Addition of Pb^{2+} beyond this concentration will result in precipitation of PbI_2. The $PbCl_2$ will not begin to precipitate until the solution is saturated with respect to $PbCl_2$, and this will not occur until

$$[Pb^{2+}] = \frac{2.4 \times 10^{-4}}{[Cl^-]^2} = \frac{2.4 \times 10^{-4}}{(0.01)^2} = 2.4 \text{ M}$$

What will $[I^-]$ be when $[Pb^{2+}] = 2.4$ M? Substituting into the K_{sp} for PbI_2 we find

$$[I^-]^2 = \frac{2.4 \times 10^{-8}}{[Pb^{2+}]} = \frac{2.4 \times 10^{-8}}{(2.4)} = 1.0 \times 10^{-8}$$

$$[I^-] = 1.0 \times 10^{-4}$$

So, at a $[Pb^{2+}]$ of 2.4 M all of the chloride ion (0.01 M) remains in solution while all but 0.0001 M of the I^- has precipitated. At a $[Pb^{2+}]$ higher than 2.4 M, $PbCl_2$ will precipitate.

PROBLEMS *Category I*

1. The K_{sp} of $ZnCO_3$ is 3.0×10^{-8}. Find the molar concentration of Zn^{2+} ion in a saturated $ZnCO_3$ solution.

2. The K_{sp} of $BaCrO_4$ is 3.0×10^{-10}. Find the molar concentration of CrO_4^{2-} ion in a saturated solution of $BaCrO_4$.

3. The K_{sp} of Ag_2CO_3 is 6.2×10^{-12}. Find the Ag^+ concentration in a saturated Ag_2CO_3 solution.

4. It is found that exactly 4.0×10^{-5} mole of strontium carbonate, $SrCO_3$, will dissolve per liter of solution. Find the K_{sp} of $SrCO_3$.

5. If the $[OH^-]$ in a saturated $Mg(OH)_2$ solution is found to be 4.0×10^{-4}, what is the K_{sp} of $Mg(OH)_2$?

Category II

6. If the K_{sp} of $Mg(OH)_2$ is 3.2×10^{-11}, what is the maximum Mg^{2+} ion concentration possible in a solution 0.001 M in OH^-?

7. The K_{sp} of $CaSO_4$ is 6.4×10^{-5}. What is the maximum SO_4^{2-} concentration possible in a 0.50 M $Ca(NO_3)_2$ solution?

8. The K_{sp} of BaF_2 is 1.7×10^{-6}. How many grams of NaF could be added to 100 ml of a 0.01 M $Ba(NO_3)_2$ solution without precipitating BaF_2?

9. The K_{sp} of HgS is 1.0×10^{-50}. If there was 1.0 gram of $HgCl_2$ in a reservoir containing 10,000 liters of water, how many moles of S^{2-} could be present in the water without precipitating HgS?

Category III

10. The K_{sp} of silver thiocyanate, AgSCN, is 1.0×10^{-12}. If 500 ml of 0.001 M $AgNO_3$ is mixed with 500 ml of 0.001 M NaSCN, will AgSCN precipitate? What is the ion product?

11. The K_{sp} of silver bromate, $AgBrO_3$, is 5.0×10^{-5}. If 500 ml of 0.001 M $AgNO_3$ is mixed with 500 ml of 0.001 M $NaBrO_3$, what is the ion product and will $AgBrO_3$ precipitate?

12. Refer to Problem 9. If 10.0 ml of 0.01 M $HgCl_2$ were added to 1000 liters of water containing 0.001 mole of S^{2-}, what would the ion product $[H^{2+}] \times [S^{2-}]$ be and would HgS precipitate?

13. A chemist has a solution containing Br^-, Cl^-, and I^-. He wishes to precipitate most of the iodide ion but leave the bromide and chloride ions in solution. K_{sp}'s of AgCl, AgBr, and AgI are 1.0×10^{-10}, 5.0×10^{-13}, and 1.0×10^{-16}, respectively. If the solution is 0.10 M in Cl^-, Br^-, and I^-, show (quantitatively) how he could do this.

14. The K_{sp}'s of Cu_2S, MnS, NiS, and PbS are 1.0×10^{-46}, 1.4×10^{-15}, 1.4×10^{-24}, and 4.2×10^{-28}. If a solution contained 1.0 mole per liter of Cu^+, Ni^{2+}, Mn^{2+}, and Pb^{2+}, which of the metal ions would precipitate if the solution were made
(a) 0.01 M in S^{2-} ion?
(b) 1.0×10^{-6} M in S^{2-} ion?
(c) 1.0×10^{-16} M in S^{2-} ion?
(d) 1.0×10^{-26} M in S^{2-} ion?
(e) 1.0×10^{-36} M in S^{2-} ion?

What would the S^{2-} concentration have to be in order to precipitate most of the Pb^{2+} but leave all of the Ni^{2+} in solution?

General Problems

15. A saturated solution of Ag_3AsO_4 was found to contain 1.0×10^{-6} mole of AsO_4^{3-} ions and 3.0×10^{-6} mole of Ag^+ ions per liter of solution. What is K_{sp} for Ag_3AsO_4?

16. To a solution containing 0.10 mole of Pb^{2+}, F^- was added. The final volume was 1.0 liter. How much F^- remained in the solution if 9.0×10^{-2} mole of PbF_2 precipitated? K_{sp} $(PbF_2) = 4.0 \times 10^{-8}$.

17. A solution is saturated with respect to both $BaSO_4$ ($K_{sp} = 1.1 \times 10^{-10}$) and BaF_2 ($K_{sp} = 1.7 \times 10^{-6}$). Find $[SO_4{}^{2-}]$ in the solution if $[F^-] = 4.1 \times 10^{-2}$ M.

18. A solution 0.027 M in $SO_4{}^{2-}$ is added to 300 ml of a solution 0.010 M in Ca^{2+}. After addition of 200 ml of the $SO_4{}^{2-}$ solution the first trace of $CaSO_4$ precipitates. Estimate the K_{sp} of $CaSO_4$.

19. A solution is originally 0.01 M in acetate ion, $C_2H_3O_2{}^-$, and 0.01 M in sulfate ion, $SO_4{}^{2-}$. Solid silver nitrate is added until $[Ag^+] = 4.0 \times 10^{-2}$ in the solution. Find $[C_2H_3O_2{}^-]$ and $[SO_4{}^{2-}]$ after the $AgNO_3$ is added.

11 ACID-BASE EQUILIBRIA

The two most widely used concepts of acids and bases are those of Bronsted-Lowry and of Arrhenius. According to the former, an acid is any proton donor and a base any proton acceptor. In the latter, an acid is any substance yielding H^+ in water and a base is any substance yielding OH^-. The two concepts are completely compatible with the latter being limited to aqueous systems. Water, itself, is both an acid and a base since it ionizes slightly to give H^+ (actually hydrated H^+, H_3O^+, called the hydronium ion) and OH^-: $H_2O \rightleftarrows H^+ + OH^-$. In pure H_2O the number of H^+ ions must equal the number of OH^- ions, and both $[H^+]$ and $[OH^-]$ are equal to 1.0×10^{-7} M (neutral). If, in an aqueous solution, $[H^+]$ is greater than $[OH^-]$, the solution is acidic and if $[OH^-]$ exceeds $[H^+]$, the solution is basic. In this chapter we will examine five categories of acid-base problems.

Problem Categories

I   The pH concept.
II Weak acids and bases.
III Hydrolysis of salts.
IV Acid-base reactions.
V   Buffer systems.

THE pH CONCEPT   CATEGORY I

The ionization of water is reversible and may be defined by an equilibrium constant:

$$H_2O \rightleftarrows H^+ + OH^-$$

$$\frac{[H^+]\,[OH^-]}{[H_2O]} = K = 1.8 \times 10^{-16}$$

This is a very small equilibrium (or "ionization") constant, and, in fact, shows that only one out of 555 million H_2O molecules is ionized at equilibrium. The ionization constant expression for water can be abbreviated. In the denominator we have $[H_2O]$, moles per liter of water. How many moles of water are there per liter of pure water? Since 1.0 liter of water weighs 1000 grams, dividing by MW of H_2O (18.0), we find there are 55.6 moles per liter of water. That number remains constant at a given temperature (55.6 at 25°C). Since K is constant and $[H_2O]$ is constant, why not combine the two in the equilibrium constant expression?

$$\frac{[H^+] \ [OH^-]}{(55.6)} = 1.8 \times 10^{-16} ; [H^+] \ [OH^-] = 1.0 \times 10^{-14}$$

The "new" constant, 1.0×10^{-14}, is called the water ionization constant, Kw. In pure water, or in dilute aqueous solutions, $[H^+] \times [OH^-]$ must equal 1.0×10^{-14}. Thus, in pure water $[H^+] = [OH^-] = 1.0 \times 10^{-7}$. If $[H^+]$ were 1.0×10^{-6}, then $[OH^-]$ would be 1.0×10^{-8}, or if $[H^+] = 1.0 \times 10^{-3}$, $[OH^-] = 1.0 \times 10^{-11}$, and so on.

The pH of a solution is simply a shorthand method of describing $[H^+]$ or $[OH^-]$ in the solution. It is defined as

$$pH = -\log \ [H^+] = \log \ \frac{1}{[H^+]}$$

So, if $[H^+] = 10^{-7}$, $\log 10^{-7} = -7.0$, and $-(-7.0) = 7.0$. The pH of a neutral solution is 7.0. Similarly, if $[H^+] = 10^{-5}$, the pH is 5.0; if $[H^+] = 10^{-9}$, the pH is 9.0, etc. Solutions with a pH less than 7 are acidic and solutions with a pH greater than 7 are basic.

Example 1.

If $[H^+]$ in a solution is 1.0×10^{-10} M, find $[OH^-]$ and the pH of the solution.

Solution.

Since $[H^+] \ [OH^-] = 1.0 \times 10^{-14}$, anytime $[H^+]$ is known we can find $[OH^-]$, and vice versa. Thus

$$[OH^-] = \frac{1.0 \times 10^{-14}}{[H^+]} = \frac{1.0 \times 10^{-14}}{1.0 \times 10^{-10}} = 1.0 \times 10^{-4}$$

$$pH = -\log (1.0 \times 10^{-10}) = -(\log 1.0 + \log 10^{-10})$$
$$= -(0 + -10.0) = 10.0$$

Example 2.

If $[H^+] = 2.0 \times 10^{-5}$, find $[OH^-]$ and pH.

Solution.

$$[OH^-] = \frac{1.0 \times 10^{-14}}{[H^+]} = \frac{1.0 \times 10^{-14}}{2.0 \times 10^{-5}} = 5.0 \times 10^{-10}$$

$$pH = -\log (2.0 \times 10^{-5}) = -(\log 2.0 + \log 10^{-5})$$
$$= -(0.30 + -5.0) = 4.70$$

There is a brief discussion of manipulating logarithms in Appendix I. A quick way to find pH from $[H^+]$ is as follows. If $[H^+] = A \times 10^{-B}$, then pH = B − log A. So, if $[H^+] = 2.0 \times 10^{-5}$, pH = 5 − log 2.0 = 5.0 − 0.30 = 4.70.

Example 3.

If $[OH^-] = 5.2 \times 10^{-3}$, what is the pH of the solution?

Solution.

$$[H^+] = \frac{1.0 \times 10^{-14}}{[OH^-]} = \frac{1.0 \times 10^{-14}}{5.2 \times 10^{-3}} = 1.9 \times 10^{-12}$$

$$pH = 12 - \log 1.9 = 12 - 0.28 = 11.72$$

Example 4.

The pH of a solution is 5.60. Find $[H^+]$ and $[OH^-]$.

Solution.

Probably the easiest way to convert from pH to $[H^+]$ is to work backwards using the "quick" method. For a pH of 5.60,

$$A \times 10^{-6}$$

Since 6 − log A = 5.60, the log of A must be 0.40. Consulting a log table we find that the log of 2.51 is 0.40, so

$$[H^+] = 2.51 \times 10^{-6}$$

Using Kw we find

$$[OH^-] = \frac{1.0 \times 10^{-14}}{2.51 \times 10^{-5}} = 3.98 \times 10^{-10}$$

Some explanation is in order here. If pH = 5.60, why must $[H^+]$ be $A \times 10^{-6}$? For any $[H^+] = A \times 10^{-B}$ we can find the pH from (B − log A), and when log A is a number between 0 and 1.0, B must be 6 in order to arrive at pH 5.60. Similarly, for pH 9.23, B must be 10, for pH 2.82, B must be 3, and so on.

WEAK ACIDS AND BASES   CATEGORY II

The only common strong acids are $HClO_4$, H_2SO_4, HNO_3, HCl, HBr, and HI. The strong bases are NaOH, KOH, RbOH, CsOH, $Ca(OH)_2$, $Sr(OH)_2$, and $Ba(OH)_2$. All other acids and bases may be classified as "weak."

How weak an acid or base is may be described by its equilibrium or ionization constant. The following table lists some weak acid ionization constants for the general ionization equilibrium $HA \rightleftarrows H^+ + A^-$.

Table 11–1 Weak Acid Ionization Constants

Acid	Formula	Ka
Sulfurous	H_2SO_3	1.7×10^{-2}
Hydrogen sulfate ion	HSO_4^-	1.2×10^{-2}
Hydrofluoric	HF	7.0×10^{-4}
Nitrous	HNO_2	4.5×10^{-4}
Formic	$HCHO_2$	2.1×10^{-4}
Benzoic	$HC_7H_5O_2$	6.6×10^{-5}
Acetic	$HC_2H_3O_2$ (HAc)	1.8×10^{-5}
Carbonic	H_2CO_3	4.2×10^{-7}
Hydrogen sulfite ion	HSO_3^-	5.6×10^{-8}
Hypochlorous	HClO	3.2×10^{-8}
Boric	H_3BO_3	5.8×10^{-10}
Hydrocyanic	HCN	4.0×10^{-10}
Hydrogen carbonate ion	HCO_3^-	4.8×10^{-11}

Since the ionization constant is a measure of how far the reaction goes from left to right, one can judge qualitatively how strong or weak an acid is by the value of the ionization constant. Thus, H_2SO_3 ionizes to a fair extent and is a stronger acid than, e.g., HNO_2, and HNO_2 is stronger than HClO, which is stronger than HCN, etc. What is the weakest acid of those listed in the table? There are very few molecular weak bases. The most common is aqueous NH_3, which ionizes as $NH_3 + H_2O \rightleftarrows NH_4^+ + OH^-$. The ionization constant is 1.8×10^{-5}.

There is no generally applicable theory to account for the values of ionization constants of weak acids and bases. They must be determined experimentally. Once the values are known, they may be used in quantitative calculations.

Example 1.

Find $[H^+]$ and the pH of a solution in which $[HCN] = 1.0$ M.

Solution.

The equilibrium is $HCN \rightleftarrows H^+ + CN^-$, and the ionization constant expression is

$$\frac{[H^+]\ [CN^-]}{[HCN]} = 4.0 \times 10^{-10}$$

We know that for pure HCN, with no other source of H^+ or CN^-, the equilibrium concentration of H^+ must equal the concentration of CN^-, since for each HCN molecule that ionizes, one H^+ and one CN^- are produced. If we let $[H^+] = X$, then $[CN^-] = X$, and we can substitute into the ionization constant expression:

$$\frac{[H^+]\ [CN^-]}{[HCN]} = \frac{(X)\ (X)}{(1.0)} = 4.0 \times 10^{-10}$$

$$X^2 = 4.0 \times 10^{-10}; \ X = 2.0 \times 10^{-5}$$

So, $[H^+] = 2.0 \times 10^{-5}$, and the pH = $(5 - \log 2.0) = (5 - 0.30) = 4.70$.

Example 2.

Calculate $[H^+]$ and the pH of a solution prepared by placing 1.0 mole of HNO_2 in enough water to give 1.0 liter of solution.

Solution.

$HNO_2 \rightleftarrows H^+ + NO_2^-$; $K = 4.5 \times 10^{-4}$.

If we set $[H^+] = X$, then for pure HNO_2 solution $[NO_2^-] = X$. But what about $[HNO_2]$ at equilibrium? We start with $[HNO_2] = 1.0$, but some will ionize to form H^+ and NO_2^-. How much? For each H^+ and NO_2^- formed, one HNO_2 must ionize, so if X moles per liter of H^+ (and NO_2^-) are formed, X moles per liter of HNO_2 are used up, and the concentration of HNO_2 left at equilibrium will be $(1.0 - X)$. Now we can substitute into the ionization constant expression:

$$\frac{[H^+]\,[NO_2^-]}{[HNO_2]} = \frac{(X)\,(X)}{(1.0 - X)} = 4.5 \times 10^{-4}$$

In order to avoid using the quadratic formula, let's approximate. Assume X is small enough to have a negligible effect in the denominator, i.e., that the denominator will be ~ 1.0. Then,

$$\frac{X^2}{1.0} = 4.5 \times 10^{-4}; \text{ and } X = 2.1 \times 10^{-2}$$

The approximation was valid, especially since we know the value of K to only two significant figures. A good rule of thumb to remember is that this kind of approximation will work if the original concentration in the denominator is larger than the value of K by a factor of $\sim 10^3$. If X had been large enough to affect the denominator $(1.0 - X)$, then we would either have to use the quadratic formula or make successive approximations. The pH of the solution is $(2 - \log 2.1) = 1.68$.

Example 3.

A solution is prepared by dissolving 0.50 mole of acetic acid, $H_2C_2H_3O_2$, (abbreviated HAc) in water to give 2.0 liters solution. Then 0.50 mole of sodium acetate, NaAc, is added. Find $[H^+]$ before and after the NaAc is added.

Solution.

First, the HAc alone will ionize slightly: $HAc \rightleftarrows H^+ + Ac^-$, $K = 1.8 \times 10^{-5}$. The $[H^+]$ will be the same as $[Ac^-]$ and $[HAc]$ will be 0.50 mole/2.0 liters = 0.25 M. Neglecting X in the denominator, we can substitute into the ionization constant:

$$\frac{[H^+]\,[Ac^-]}{[HAc]} = \frac{(X)\,(X)}{(0.25)} = 1.8 \times 10^{-5}$$

$$X^2 = 4.5 \times 10^{-6}, X = [H^+] = 2.12 \times 10^{-3}$$

Second, NaAc is a salt and will ionize completely into Na^+ and Ac^-, so, in effect, we are adding Ac^- to the $HAc \rightleftarrows H^+ + Ac^-$ equilibrium. The equilibrium will shift to the left and $[H^+]$ will get smaller. No longer will $[H^+] = [Ac^-]$ since we now have a source of Ac^- other than just pure HAc. What is $[Ac^-]$? As a good approximation, we can say essentially all of the Ac^- comes from the NaAc, neglect the very small amount from HAc, and set $[Ac^-] = 0.50/2.0 \, l = 0.25$ M. The ionization constant expression for HAc must remain constant, so we can now substitute:

$$\frac{[H^+] \, [Ac^-]}{[HAc]} = \frac{[H^+] \, (0.25)}{(0.25)} = 1.8 \times 10^{-5}$$

$$[H^+] = 1.8 \times 10^{-5}$$

Sure enough, the hydrogen ion concentration decreased upon addition of Ac^-. Strictly speaking, the [HAc] would increase and be slightly greater than 0.25 M, but the increase is so slight that it can be neglected in the calculation.

Example 4.

How many moles of sodium formate, $NaCHO_2$, must be added to 250 ml of a 1.0 M $HCHO_2$ solution to give a pH of exactly 4.00?

Solution.

We know $[HCHO_2] = 1.0$, and $[H^+]$ must be 1.0×10^{-4}, so we can solve for $[CHO_2^-]$ in the ionization constant expression:

$$\frac{[H^+] \, [CHO_2^-]}{[HCHO_2]} = \frac{(1.0 \times 10^{-4}) \, [CHO_2^-]}{(1.0)} = 2.1 \times 10^{-4}$$

$$[CHO_2^-] = 2.1$$

And if the number of moles per liter must be 2.1, the number of moles that must be added to 250 ml is $2.1 \div 4 = 0.53$.

Example 5.

Find the pH of 1.0 liter of 1.0 M ammonia solution.

Solution.

The ionization is $NH_3 + H_2O \rightleftarrows NH_4^+ + OH^-$ with $K = 1.8 \times 10^{-5}$. In the pure NH_3 solution, we can set $[NH_4^+] = [OH^-] = X$, and $[NH_3] = 1.0$. Then,

$$\frac{[NH_4^+] \, [OH^-]}{[NH_3]} = \frac{(X) \, (X)}{1.0} = 1.8 \times 10^{-5}$$

$$X = [OH^-] = 4.2 \times 10^{-3}$$

Substituting into Kw,

$$[H^+] = \frac{1.0 \times 10^{-14}}{4.2 \times 10^{-3}} = 2.4 \times 10^{-12}$$

and,

$$pH = (12 - \log 2.4) = 11.62$$

There is a direct relationship between the acid and base properties of a substance. Let's use the NH_3 equilibrium to illustrate this relationship. For NH_3, as a weak base, we write

$$NH_3 + H_2O \rightleftarrows NH_4^+ + OH^-$$

and describe the ionization by the constant, $K = 1.8 \times 10^{-5}$. According to the

Bronsted-Lowry concept, NH_3 behaves as a base because it accepts a proton (from H_2O which acts as an acid by donating a proton). But, remember, we can read an equilibrium reaction both ways. What about the reverse reaction?

$$NH_4^+ + OH^- \rightleftarrows NH_3 + H_2O$$

What is the acid and what is the base? The NH_4^+ acts as an acid by donating a proton to OH^- (the base). So, while NH_3 is a base, the NH_4^+ ion is an acid. In the terminology of the Bronsted-Lowry concept, the NH_4^+ ion is called the "conjugate" acid of the base NH_3, and/or NH_3 is the "conjugate" base of the acid NH_4^+. We will explore this idea further in the next category.

HYDROLYSIS OF SALTS   CATEGORY III

The term "hydrolysis" means reaction with water. Water, itself, is simultaneously both a weak acid and a weak base with an ionization constant of 1.8×10^{-16} (see Category I). As such, water can, and does, react with other bases and acids.

One advantage of the Bronsted-Lowry concept is that it broadly defines a base as *any* proton acceptor, without being limited to hydroxide ion, OH^-. While OH^- is certainly a good proton acceptor and the equilibrium $H^+ + OH^- \rightleftarrows HOH$ lies very far to the right, other bases also have a strong tendency to accept protons. For example,

$$
\begin{aligned}
H^+ + O^{2-} &\rightleftarrows OH^- \\
H^+ + NH_3 &\rightleftarrows NH_4^+ \\
H^+ + HS^- &\rightleftarrows H_2S \\
H^+ + CN^- &\rightleftarrows HCN \\
H^+ + Ac^- &\rightleftarrows HAc \\
H^+ + F^- &\rightleftarrows HF \\
H^+ + HSO_3^- &\rightleftarrows H_2SO_3
\end{aligned}
$$

How readily the base accepts protons depends upon how "strong" the base is, or how "weak" is the "conjugate" acid that is formed. Thus, OH^- is a very, very weak acid and O^{2-} is a very strong base, but H_2SO_3 is a fairly strong acid, so HSO_3^- is a weak base.

Since H_2O can act as an acid, any base (depending upon its strngth) can accept a proton from a water molecule. Conversely, any acid (depending upon its strength) can donate a proton to, or accept an OH^- from, a water molecule.

Example 1.

Write hydrolysis reactions for the following ions: (a) CN^- (b) Ac^- (c) NH_4^+ (d) Na^+ (e) Fe^{3+} (f) Cl^-.

Solution.

(a) $CN^- + HOH \rightleftarrows HCN + OH^-$

The cyanide ion is a good base since HCN is a very weak acid. Thus, in water, the CN^- ion "sees" a source of protons, and to a small extent, at least, CN^- ions take protons away from H_2O to form some HCN molecules. Water is written HOH simply to emphasize the acid-base nature of water molecules. Note that when a base such as

CN^- takes protons away from H_2O molecules, the other product must be OH^- ions. So, as the CN^- ion hydrolyzes, OH^- ions are released into the solution and the solution becomes basic.

(b) $Ac^- + HOH \rightleftarrows HAc + OH^-$

Like CN^-, Ac^- is a base since HAc is a weak acid. The reaction proceeds further for CN^- than it does for Ac^- since CN^- is a stronger base than Ac^-, i.e., HCN is a weaker acid than HAc.

(c) $NH_4^+ + HOH \rightleftarrows NH_3 + H_3O^+$

Since NH_4^+ is a weak acid, there is a tendency for NH_4^+ ions to donate protons. H_2O can serve as a base. When this happens, H^+ (H_3O^+) is released into the solution and the solution becomes acidic.

(d)   No reaction.

Na^+ ions certainly have no tendency to combine with H^+, and there is also no tendency to combine with OH^- to form un-ionized NaOH. NaOH is a strong base and is completely ionized in solution.

(e) $Fe^{3+} + HOH \rightleftarrows FeOH^{2+} + H^+$

$Fe(OH)_3$ is a weak base and metal ions from weak bases do hydrolyze. The hydrolysis may proceed further:

$$FeOH^{2+} + HOH \rightleftarrows Fe(OH)_2^+ + H^+$$

$$Fe(OH)_2^+ + HOH \rightleftarrows Fe(OH)_3 + H^+$$

But, the first step is the most extensive. Note that metal ions that hydrolyze give acid solutions due to the release of H^+.

(f)   No reaction.

Only ions from weak acids or bases will hydrolyze. The Cl^- ion comes from the strong acid, HCl, and since HCl is completely ionized in solution, there is no tendency for Cl^- to combine with H^+ from HOH.

Example 2.

Calculate the pH of a 1.0 M NaCN solution.

Solution.

Sodium cyanide is a salt. It is formed from the weak acid HCN and the strong base NaOH. The only way a salt solution can have a pH other than 7.0 is if one (or both) of the ions hydrolyze. Salts of strong acids and strong bases (e.g., NaCl) do not hydrolyze. Salts of weak bases and strong acids (e.g., NH_4Cl) hydrolyze to give acidic solutions. Salts of strong bases and weak acids (e.g., NaCN) hydrolyze to give basic solutions. In solution NaCN exists as Na^+ and CN^-. The Na^+ ion does not hydrolyze, but the CN^- ion does:

$$CN^- + HOH \rightleftarrows HCN + OH^-$$

The hydrolysis is an equilibrium, just the reverse of neutralization, and we can write an equilibrium expression:

$$\frac{[HCN]\,[OH^-]}{[CN^-]} = K$$

Why doesn't [HOH] appear in the expression?

The hydrolysis equilibrium constant may be determined by experiment, or we can deduce it as follows. Multiply top and bottom of the K expression by $[H^+]$.

$$\frac{[HCN]\,[OH^-]}{[CN^-]} \times \frac{[H^+]}{[H^+]}$$

Multiplying and dividing by the same factor doesn't change anything, it is just multiplying something $\times$ 1. But notice that we now have $[H^+] \times [OH^-]$ in the numerator. That is Kw. If we took Kw out of the expression, we would be left with

$$\frac{[HCN]}{[CN^-]\,[H^+]}$$

which is just the ionization constant for HCN upside-down, or $1/K_a(HCN)$. We can now disregard the "fudge" factor $[H^+]/[H^+]$, but we have shown that

$$\frac{[HCN]\,[OH^-]}{[CN^-]} = \frac{Kw}{K_a(HCN)} = K_b(CN^-)$$

For the CN^- ion, $K_b = 1.0 \times 10^{-14}/4.0 \times 10^{-10} = 2.5 \times 10^{-5}$. Finally, we can solve this problem like any other equilibrium problem:

$$\begin{array}{ccc} (1.0 - X) & X & X \\ CN^- \; + HOH \rightleftarrows HCN + & OH^- \end{array}$$

and, neglecting X in the denominator,

$$\frac{[HCN]\,[OH^-]}{[CN^-]} = \frac{(X)\,(X)}{(1.0)} = 2.5 \times 10^{-5}$$

$$X = [OH^-] = 5.0 \times 10^{-3},\ [H^+] = \frac{1.0 \times 10^{-14}}{5.0 \times 10^{-3}} = 2.0 \times 10^{-12},\ \text{and}$$

$$pH = (12 - \log 2.0) = 11.70$$

Notice three things. First, the solution is definitely basic with a pH of 11.70. Second, even for a strong base ion like CN^-, hydrolysis does not occur to a *great* extent, since the equilibrium constant is only 2.5×10^{-5} for the reaction.

Third, in order to solve the problem we have calculated an equilibrium constant, K_b, for the CN^- ion acting as a base. The conjugate acid of CN^- is HCN, and we showed that $K_b(CN^-)$ was equal to $Kw/K_a(HCN)$. In fact, for any conjugate acid and base

$$\begin{array}{cc} HB \rightleftarrows H^+ + & B^- \\ \text{acid} & \text{base} \end{array}$$

the relationship between K_a and K_b is

$$K_a \times K_b = Kw$$

Example 3.

Calculate the hydrolysis constant (i.e., K_a) for the ammonium ion, NH_4^+.

Solution.

The ion hydrolyzes to give an acidic solution:

$$NH_4^+ + HOH \rightleftarrows NH_3 + H_3O^+$$

In this case, we can multiply and divide the equilibrium expression by $[OH^-]$ to show that $K_a = K_w/K_b$ (NH_3):

$$\frac{[NH_3]\ [H_3O^+]}{[NH_4^+]} \times \frac{[OH^-]}{[OH^-]} = \frac{K_w}{K_b\ (NH_3)}$$

Thus, the K_a for the NH_4^+ ion is

$$\frac{1.0 \times 10^{-14}}{1.8 \times 10^{-5}} = 5.6 \times 10^{-10}$$

Example 4.

A 1.0 M $Al(NO_3)_3$ solution is found to have a pH of 2.70. Calculate K_a for the reaction

$$Al^{3+} + HOH \rightleftarrows AlOH^{2+} + H^+$$

Solution.

For pH 2.70, the $[H^+] = A \times 10^{-3}$, log A must be 0.30, so $[H^+] = 2.0 \times 10^{-3}$. From the hydrolysis equilibrium we see that if $[H^+] = 2.0 \times 10^{-3}$, $[AlOH^{2+}]$ must also be 2.0×10^{-3}, and $[Al^{3+}]$ would be $1.0 - .002 \simeq 1.0$. So,

$$\frac{[AlOH^{2+}]\ [H^+]}{[Al^{3+}]} = \frac{(2.0 \times 10^{-3})\ (2.0 \times 10^{-3})}{(1.0)} = 4.0 \times 10^{-6}$$

CATEGORY IV   ACID-BASE REACTIONS

Strong acids and bases are those that ionize completely, or nearly completely, in solution. Weak acids and bases are those that do not. For example, HNO_3 is a strong acid because the reaction $HNO_3 \rightarrow H^+ + NO_3^+$ is essentially irreversible, i.e., the reaction goes completely from left to right. On the other hand, HNO_2 is a weak acid because the reaction $HNO_2 \leftrightarrows H^+ + NO_2^-$ is reversible and equilibrium exists when the HNO_2 is only slightly ionized (the ionization constant is 4.5×10^{-4}).

Any acid, strong or weak, will react with any strong or weak base. Acid-base reactions are best represented by *net ionic equations*. In a net ionic equation all species are written as they exist in solution and only those species that undergo chemical change are included in the equation. For example, let's examine the reaction of HCl with NaOH. Writing the reaction as $HCl + NaOH \rightarrow NaCl + H_2O$ is misleading. In aqueous solution, HCl, NaOH, and NaCl exist as ions, not as molecules. It would be better to write:

$$H^+ + Cl^- + Na^+ + OH^- \rightarrow Na^+ + Cl^- + H_2O$$

The water does exist primarily as undissociated molecules. Note that we now have Na^+ and Cl^- ions on both sides of the equation, and these do not undergo any type of chemical change. They can be cancelled out of the equation, leaving

$$H^+ + OH^- \rightarrow H_2O$$

This is the net ionic equation for the reaction of HCl with NaOH, or, in fact, for the reaction of any strong acid with any strong base. For reactions of strong acids with weak bases, or vice-versa, net ionic equations would be written, for example, as

$$H^+(aq) + NH_3(aq) \rightarrow NH_4^+(aq)$$

$$HNO_2(aq) + OH^-(aq) \rightarrow NO_2^-(aq) + H_2O$$

since weak acids and bases exist primarily undissociated in solution and do undergo change as the salt is formed in the neutralization. How would you write the net ionic equation for neutralization of a weak acid with a weak base?

Example 1.

Write net ionic equations for the following acid-base reactions (in aqueous solution: (a) KOH + HCl (b) HCN + KOH (c) HNO_3 + NH_3 (d) NH_3 + HCN.

Solution.

(a) strong base — strong acid

$$H^+ + OH^- \rightarrow H_2O$$

The K^+ and Cl^- ions are unchanged and are not included.

(b) weak acid — strong base

$$HCN + OH^- \rightarrow CN^- + H_2O$$

The weak acid HCN is undissociated, but the salt KCN is ionized. The K^+ ions are unchanged and do not appear in the equation.

(c) strong acid — weak base

$$H^+ + NH_3 \rightarrow NH_4^+$$

The NO_3^- ions are unchanged in the reaction.

(d) weak acid — weak base

$$HCN + NH_3 \rightleftarrows NH_4^+ + CN^-$$

Both HCN and NH_3 are primarily un-ionized, but NH_4CN is ionized, as are (almost) all salts in aqueous solution.

Now, let us look at some quantitative examples of acid-base (neutralization) reactions.

Example 2.

Sodium hydroxide is added to one liter of 0.100 M HCl solution. Calculate $[H^+]$ and pH after the following amounts of NaOH have been added: (a) none (b) 0.05 mole (c) 0.09 mole (d) 0.099 mole (e) 0.10 mole (f) 0.11 mole.

Solution.

(a) Since HCl is a strong acid, a solution 0.10 M in HCl is 0.10 M in H^+. So, $[H^+] = 10^{-1}$ and pH = 1.0.

(b) The net reaction is $H^+ + OH^- \rightarrow H_2O$. Addition of 0.05 mole of OH^- will neutralize only half of the 0.100 mole of H^+ present, leaving 0.05 mole of H^+ unreacted. So. $[H^+] = 0.05 = 5 \times 10^{-2}$, and pH = 2 - log 5 = 2 - 0.70 = 1.30.

(c) The amount of H^+ remaining will be 0.100 - 0.09 = 0.01, so $[H^+] = 10^{-2}$ and the pH is 2.0.

(d) The amount of H^+ remaining will be 0.100 - 0.099 = 0.001, so $[H^+] = 10^{-3}$ and the pH is 3.0.

(e) This is the point where all H^+ from the HCl has been neutralized and we have, in effect, a solution of NaCl. Any H^+ in the solution comes from the H_2O, so $[H^+] = 10^{-7}$ and the pH is 7.0.

(f) We now have 0.01 mole of OH^- in excess. It is as if we had added 0.01 mole of OH^- to a NaCl solution. So, at this point $[OH^-] = 0.01 = 10^{-2}$, so $[H^+] = 10^{-12}$ and the pH is 12.0.

Notice how drastically the pH of the solution changes near the point of complete neutralization, from pH 3.0 in part (d) to 12.0 in part (f). This is the basis for acid-base titration. The large pH change near the point of neutralization is usually detected by means of a suitable indicator.

Example 3.

Exactly 10.0 grams of KOH are added to 100 ml of water. How many moles of H_2CO_3 would be needed to neutralize the KOH?

Solution.

The net reaction is $H_2CO_3 + 2 OH^- \rightleftarrows CO_3^{2-} + 2 H_2O$, so it will take one mole of H_2CO_3 for every two moles of KOH to affect neutralization. The number of moles of KOH is

$$\text{no. of moles} = \frac{wt}{\text{formula wt}} = \frac{10.0 \text{ g}}{56.1 \text{ g}} = 0.18 \text{ mole}$$

Therefore, it would require 0.09 mole of H_2CO_3. The 100 ml of water is unimportant in this problem. We needed to know the number of moles of H_2CO_3 to neutralize 10.0 g of KOH and it made no difference whether that 10.0 g of KOH was in 100 ml of water or a barrel of water.

Example 4.

How many mls of 0.10 M HNO_3 are needed to neutralize 25.0 ml of 0.50 M NH_3 solution?

Solution.

The net reaction is $H^+ + NH_3 \rightleftarrows NH_4^+$, and one mole of HNO_3 is required for each mole of NH_3. In this problem the volume is important. Since we know the molarity of the NH_3 solution and we know the volume, we can find the number of moles of NH_3:

$$\text{moles} = M \times \text{liters} = 0.50 \text{ M} \times 0.025 \text{ } \ell = 0.0125 \text{ mole}$$

Thus, we now know the number of moles of HNO_3 required (0.0125) and we know the molarity of the HNO_3, so we can find the volume needed:

$$\text{liters} = \frac{\text{moles}}{M} = \frac{0.0125}{0.10} = 0.125 \text{ } \ell = 125.0 \text{ ml}$$

Example 5.

How many mls of 0.20 M H_2SO_4 are required to neutralize 1.0 gram of NaOH?

Solution.

Since H_2SO_4 is a diprotic (two proton) acid, one mole of H_2SO_4 will neutralize two moles of NaOH. The number of moles of NaOH is wt/formula wt = 1.0 g/40.0 = 0.025 mole. So, the number of moles of H_2SO_4 needed will be one half of that, or 0.0125 mole. Knowing the molarity of the H_2SO_4 we find

$$\text{liters} = \text{moles}/M = 0.0125 \text{ mole}/0.20 \text{ mole per liter} = 0.0625 \ \ell = 62.5 \text{ ml}$$

BUFFER SYSTEMS   CATEGORY V

Buffer systems are solutions that resist changes in pH. Human blood is an excellent example of a buffer system. Buffers consist of mixtures of a weak acid and a salt of the weak acid or a weak base and a salt of the weak base. The mechanism of buffer solutions is best illustrated by example.

Example 1.

A buffer solution is prepared by adding enough HAc and NaAc to 1.0 liter of solution to give [HAc] = 1.0 and [Ac⁻] = 1.0.
   (a)   Find the pH of the solution.
   (b)   Find the pH of the solution after adding 0.10 mole of NaOH.
   (c)   Find the pH of the solution after adding 0.10 mole of HCl.

Solution.

   (a)   Substituting into the ionization constant expression for HAc we get

$$\frac{[H^+] \ [Ac^-]}{[HAc]} = \frac{[H^+] \ (1.0)}{(1.0)} = 1.8 \times 10^{-5}$$

   $[H^+] = 1.8 \times 10^{-5}$, and pH = (5 − log 1.8) = 4.75

   (b)   Addition of NaOH will neutralize part (0.10 mole) of the HAc and form 0.10 mole of Ac⁻:

$$HAc + OH^- \rightleftarrows Ac^- + H_2O$$

   Thus, after reaction, [HAc] will decrease from 1.0 M to 0.90 M and [Ac⁻] will increase from 1.0 M to 1.10 M. Substituting the new values into the K expression, we find

$$\frac{[H^+] \ [Ac^-]}{[HAc]} = \frac{[H^+] \ (1.10)}{(0.90)} = 1.8 \times 10^{-5}$$

   $[H^+] = 1.5 \times 10^{-5}$, and pH = (5 − log 1.5) = 4.82

   So, addition of 0.10 mole of the strong base, NaOH, changes the pH from 4.75 to 4.82, i.e., there is very little change. What would have happened if this were not a buffer system? Imagine adding 1.8×10^{-5} moles of HCl to a liter of water so that $[H^+] = 1.8 \times 10^{-5}$ and the pH was 4.75. What would happen if 0.10 mole of NaOH were added to that solution? The .000018 mole of HCl would be quickly neutralized and it would be like adding 0.10 mole of NaOH to a liter of water! After the NaOH was added, the $[OH^-]$ would be 0.10 M and the pH would be 13.0. In the un-buffered solution the pH would change from 4.75 to 13.0, but in the buffer solution the pH change is only from 4.75 to 4.82.

(c) If we add a strong acid such as HCl to the buffer, what will happen? Is there anything in the buffer solution that will use up (react with) the H^+ ions? Indeed there is. The equilibrium $H^+ + Ac^- \rightleftarrows HAc$ lies far to the right, and the buffer solution contains an excess of Ac^- ions. As the 0.10 mole of H^+ is added, the 1.0 mole of Ac^- ions will effectively consume all of the hydrogen ions, forming 0.10 mole of HAc and using up 0.10 mole of Ac^-. The new concentrations of HAc and Ac^- will be 1.10 M and 0.90 M, respectively. We can now substitute these values into the K expression:

$$\frac{[H^+] [Ac^-]}{[HAc]} = \frac{[H^+] (0.90)}{(1.10)} = 1.8 \times 10^{-5}$$

$$[H^+] = 2.2 \times 10^{-5}, \text{ and pH} = (5 - \log 2.2) = 4.66.$$

So, on addition of 0.10 mole of strong acid, the buffer changes pH from 4.75 to 4.66. What would happen in an un-buffered solution?

Example 2.

A solution is prepared by mixing 1.60 moles of NH_3, 1.20 moles of NH_4Cl, and 0.40 mole of HCl in water to give 1.00 liter of solution. Calculate the pH of the solution.

Solution.

One may think of this as a solution originally 1.60 M in NH_3 and 1.20 M in NH_4^+ (NH_4Cl), to which is added 0.40 mole per liter of HCl. It is then recognizable as a buffer problem similar to Example 1. When the HCl is added, some of the NH_3 will be neutralized:

$$NH_3 + H^+ \rightarrow NH_4^+$$

In fact, 0.40 mole per liter of the NH_3 will be consumed and an additional 0.40 mole per liter of NH_4^+ will be formed. When equilibrium is established we will have $[NH_3] = 1.20$ and $[NH_4^+] = 1.60$ M, and we can substitute these values into the ionization constant for NH_3.

$$\frac{[NH_4^+] [OH^-]}{[NH_3]} = \frac{(1.60) [OH^-]}{(1.20)} = 1.8 \times 10^{-5}$$

$$[OH^-] = 1.35 \times 10^{-5}; [H^+] = \frac{1.0 \times 10^{-14}}{1.35 \times 10^{-5}} = 7.4 \times 10^{-10}$$

$$pH = (10 - \log 7.4) = 9.13$$

Find the pH "before" the HCl is added. Answer: 9.38.

Category I

1. Find $[OH^-]$ for solutions in which $[H^+]$ is (a) 2.6×10^{-4} (b) 8.72×10^{-9} (c) 1.00 (d) 4.3×10^{-6} (e) 6.66×10^{-12}

2. Find the pH of the solutions in Problem 1.

3. Find the pH of the following solutions: (a) 0.012 M NaOH (b) 0.20 M NaCl (c) 0.0075 M HNO_3 (d) 0.50 gram KOH in 5.0 liters of water (e) 10.0 ml 0.25 M HCl diluted with 290.0 ml of water.

4. Find $[H^+]$ for solutions of the following pH: (a) 4.70 (b) 9.32 (c) –1.0 (d) 7.30 (e) 5.19.

5. Complete the following table.

Solution	$[H^+]$	$[OH^-]$	pH
a		2.0×10^{-4}	
b			8.80
c	2.0×10^{-4}		
d	6.45×10^{-5}		
e			6.45

Category II

6. Find the pH of a 1.0 M HClO Solution ($K = 3.2 \times 10^{-8}$).

7. What is $[H^+]$ in a solution prepared by dissolving 0.50 mole of hydrofluoric acid, HF, in enough water to give 250 ml of solution? ($K = 7.0 \times 10^{-4}$).

8. Find the pH of a solution 0.50 M in benzoic acid, $HC_7H_5O_2$, and 0.20 M in sodium benzoate, $NaC_7H_5O_2$.

9. A 1.0 M solution of propionic acid, $HC_3H_5O_2$, has a pH of 2.42. What is the ionization constant of $HC_3H_5O_2$?

10. Find $[H^+]$ in a 0.50 M $NaHSO_4$ solution. (K (HSO_4^-) = 1.2×10^{-2}).

11. Calculate $[Ac^-]$ in a solution 1.0 M in HAc and 0.50 M in HCl.

Category III

12. In aqueous solution which of the following salts would not hydrolyze, would hydrolyze to give an acid solution, or would hydrolyze to give a basic solution? (a) $CrCl_3$ (b) KClO (c) NaF (d) NaBr (e) NH_4NO_3 (f) KNO_3 (g) NH_4CN.

13. Find the pH of a 2.0 M NH_4NO_3 solution.

14. The pH of a 0.10 M $AgNO_3$ solution is found to be 4.85. What is the hydrolysis constant (or Ka) for the Ag^+ ion?

Category IV

15. How many moles of NaOH are required to neutralize 100.0 ml of 0.25 M HNO_3?

16. How many moles of NaOH are required to completely react with 100.0 ml of 0.50 M H_3PO_4?

17. A 0.10 M HCl solution is added to 50.0 ml of 0.10 M NaOH. Assume volumes are additive and complete the following table.

mls HCl added	$[OH^-]$	$[H^+]$	pH
0			
20.0			
40.0			
49.0			
49.9			
50.0			
50.1			
51.0			
60.0			

18. The irritant in an ant "bite" is formic acid, $HCHO_2$. Why does "baking soda" relieve the effects of an ant "bite"?

19. How many grams of calcium hydroxide, $Ca(OH)_2$, would be needed to neutralize 10.0 grams of hydrobromic acid, HBr?

Category V

20. Which of the following combinations would give a buffer solution? (a) HNO_3 + NH_4NO_3 (b) $NaNO_2$ + NaOH (c) HNO_2 + $NaNO_2$ (d) NH_4NO_3 + NH_4OH (e) HCl + KOH.

21. How many grams of sodium acetate, NaAc, must be added to 250 ml of a 0.50 M HAc solution to give a buffer solution of pH 4.05?

22. A buffer solution contains 0.20 mole of NH_4OH and 0.20 mole NH_4Cl in 100 ml of solution. (a) What is the pH of the solution? (b) What is the pH after addition of 0.05 mole of HCl? (c) What is the pH after addition of 0.05 mole of NaOH?

23. Find $[H^+]$ in a solution prepared by mixing 1.0 mole of HF, 1.0 mole of NaF, and 0.20 mole of NaOH in one liter of solution.

General Problems

24. An aqueous solution 0.50 M in methyl amine, CH_3NH_2, is found to have a pH of 12.20. Write the reaction of CH_3NH_2 acting as a base and determine Kb.

25. Determine K for the following equilibria:
(a) $CH_3NH_3^+ + H_2O \rightleftarrows CH_3NH_2 + H_3O^+$ (see Prob. 24) (or $CH_3NH_3^+ \rightleftarrows CH_3NH_2 + H^+$)
(b) $HClO + OH^- \rightleftarrows ClO^- + H_2O$
(c) $NO_2^- + H_2O \rightleftarrows HNO_2 + OH^-$

26. An aqueous solution of formic acid, $HCHO_2$, and sodium formate, $NaCHO_2$, has a pH of 3.38. What is the ratio $[CHO_2^-]/[HCHO_2]$?

27. See Category III, Example 4. In which solution is $[H^+]$ greater:
(a) 1.0 mole of $Al(NO_3)_3$ in 1.0 liter, or
(b) 1.0 mole HClO in 1.0 liter?

28. A buffer solution is 0.50 M in NH_3 and 0.50 M in NH_4Cl. How many moles of NaOH must be added to 1.00 liter of this solution in order to increase the pH by 1.00 pH unit? (Assume no volume change.)

29. The HCO_3^- ion can act as an acid or a base. For $H_2CO_3 \rightleftarrows H^+ + HCO_3^-$, $Ka = 4.2 \times 10^{-7}$ For $HCO_3^- \rightleftarrows H^+ + CO_3^{2-}$, $Ka = 5.6 \times 10^{-11}$. Determine whether a solution 1.0 M in HCO_3^- would be acidic, basic, or neutral.

30. When 0.50 mole of sodium butyrate, $NaC_4H_7O_2$, and 0.25 mole of HCl are dissolved in water to give 500 ml of solution, the pH of the solution is found to be 4.82. Estimate Ka for butyric acid, $HC_4H_7O_2$.

12 ELECTROCHEMISTRY

Traditionally the term "electrochemistry" referred to the connection between chemistry and electricity. Any chemical process that involves a transfer of electrons can be loosely termed electrochemistry. *Oxidation* is the process of losing electrons and *reduction* refers to the gaining of electrons. The *oxidation state*, or oxidation number, is the charge that an atom appears to have, whether alone or in combination with other atoms. Oxidation states may be assigned by following a few simple rules. (1) The oxidation state of an atom in an elementary substance is zero; (2) the oxidation state of a monatomic ion is the charge on the ion; (3) in combination, the oxidation state of O is almost always -2 and the oxidation state of H is almost always $+1$; (4) the sum of the oxidation states of all the elements in a species must add up to the charge on that species.

Thus, in a reaction such as the oxidation of iron, $2\ Fe(s) + 3/2\ O_2(g) \rightarrow Fe_2O_3(s)$, the elemental iron and oxygen have oxidation states of 0. In Fe_2O_3 the oxidation states of Fe and O are $+3$ and -2, respectively. The iron loses electrons and is oxidized. The oxygen gains electrons and is reduced. In the uncharged molecule HNO_2, if we assign each O atom an oxidation state of -2 and the H atom an oxidation state of $+1$, the N atom must have an oxidation state of $+3$ in order that the charges all balance. In the ion $CrO_4{}^{2-}$, there are a total of $8-$ charges from the four O atoms; therefore, the oxidation state of Cr must be $+6$ in order that the $CrO_4{}^{2-}$ ion have a charge of $2-$.

In this chapter we will examine three categories of problems dealing with oxidation-reduction phenomena.

Problem Categories

  I   Balancing oxidation-reduction equations.
  II Electrochemical cells and standard voltages.
 III Spontaneity and extent of oxidation-reduction reactions.

CATEGORY I   BALANCING REDOX EQUATIONS

For every oxidation there must be a simultaneous reduction. One substance cannot lose electrons without another substance gaining electrons. In chemical reactions in which electrons are transferred, the number of electrons lost by the substance being oxidized must equal the number of electrons gained by the substance being reduced. An important part of understanding chemical stoichiometry is the ability to write balanced equations for oxidation-reduction reactions. There are various techniques, or tricks of the trade, for balancing "redox" reactions. In aqueous solutions, H_2O, H^+, or OH^- are often involved as part of an oxidation-reduction reaction, and the best balancing technique is the "half-reaction method" that enables one to both balance and complete a redox reaction. The method involves six steps, which may sound complicated at first, but with a little practice you will find them to be quite simple. The steps are (1) separate the net reaction expression into half-reactions; (2) balance all atoms other than O and H; (3) balance O and H using H_2O and H^+ in acid and OH^- and H_2O in base; (4) balance the charge by adding electrons; (5) multiply the half-reactions by whatever numbers needed to given an identical number of electrons in each half-reaction; (6) add the half-reactions and cancel the electrons and any H^+, H_2O, or OH^- which appear on both sides of the equation.

Example 1.

It is observed that copper metal dissolves in nitric acid, producing Cu^{2+} ions and NO gas. Write a complete, balanced equation for the reaction.

Solution.

The net reaction expression is

$$Cu(s) + NO_3^-(aq) \rightarrow Cu^{2+}(aq) + NO(g)$$

Following the six steps:
(1) Separate into half-reactions.

$$Cu \rightarrow Cu^{2+}$$

$$NO_3^- \rightarrow NO$$

(2) Balance all atoms other than O and H.
Not needed in this example. There is one copper on each side of the first half-reaction and one nitrogen on each side of the second half-reaction. Had the second half-reaction been $NO_3^- \rightarrow N_2O$, we would complete this step by making 2 $NO_3^- \rightarrow N_2O$.
(3) Balance O and H using H_2O and H^+ in acid.

$$Cu \rightarrow Cu^{2+}$$

$$4\,H^+ + NO_3^- \rightarrow NO + 2\,H_2O$$

Note that H_2O molecules are added to balance O's, then H^+ are added to the other side to balance the H's.
(4) Balance the charge by adding electrons.

$$Cu \rightarrow Cu^{2+}$$

The way it stands, the charge on the left is zero and the charge on the right is 2+. We can balance the charge by adding 2 "minus", i.e., two electrons, to the right side.

$$Cu \rightarrow Cu^{2+} + 2\,e^-$$

For the other half-reaction:

$$4\,H^+ + NO_3^- \rightarrow NO + 2\,H_2O$$

The total charge on the left is (4+ and 1− = 3+) and is zero on the right side. We can balance the charge by adding 3 e⁻ to the left side:

$$3\,e^- + 4\,H^+ + NO_3^- \rightarrow NO + 2\,H_2O$$

(5) Multiply the half-reactions by whatever numbers are needed to given an identical number of electrons in each half-reaction.

We now have

$$Cu \rightarrow Cu^{2+} + 2\,e^-$$

$$3e^- + 4\,H^+ + NO_3^- \rightarrow NO + 2\,H_2O$$

To get the same number of electrons in each half-reaction, we must multiply the top reaction by 3 and the bottom reaction by 2, giving six electrons in each. Note that the *entire* half-reaction is multiplied, not just the electrons.

$$3\,Cu \rightarrow 3\,Cu^{2+} + 6\,e^-$$

$$6\,e^- + 8\,H^+ + 2\,NO_3^- \rightarrow 2\,NO + 4\,H_2O$$

(6) Add the two half-reactions and cancel the e⁻'s and any H⁺, H₂O, or OH⁻ appearing on both sides. Adding, we get

$$3\,Cu + 6\,e^- + 8\,H^+ + 2\,NO_3^- \rightarrow 3\,Cu^{2+} + 6\,e^- + 2\,NO + 4\,H_2O$$

and in this example, only the e⁻'s cancel. The final complete and balanced equation is

$$3\,Cu(s) + 8\,H^+(aq) + 2\,NO_3^-(aq) \rightarrow 3\,Cu^{2+}(aq) + 2\,NO(g) + 4\,H_2O(l)$$

The final equation may be checked by seeing that all atoms balance and that the net charges balance. In our example, the net charge on the left is 8+ and 2− = 6+, and on the right it is 3 × 2+ = 6+. Everything checks.

While they may be considered dubious advantages, notice two aspects of the half-reaction method. First, we didn't actually *have* to know what was oxidized and what was reduced, and second, we didn't *have* to know the oxidation states of any of the atoms involved in the reaction.

Example 2.

Sulfur dioxide reacts with NO_3^- in acid solution to produce SO_4^{2-} and N_2O. Write a complete, balanced equation for the reaction.

Solution.

(1) $\qquad\qquad\qquad\qquad SO_2 \rightarrow SO_4^{2-}$

$$NO_3^- \rightarrow N_2O$$

(2) $\qquad\qquad\qquad\qquad SO_2 \rightarrow SO_4^{2-}$

$$2\,NO_3^- \rightarrow N_2O$$

(3) $$2 H_2O + SO_2 \rightarrow SO_4^{2-} + 4 H^+$$

$$10 H^+ + 2 NO_3^- \rightarrow N_2O + 5 H_2O$$

(4) $$2 H_2O + SO_2 \rightarrow SO_4^{2-} + 4 H^+ + 2 e^-$$

$$8 e^- + 10 H^+ + 2 NO_3^- \rightarrow N_2O + 5 H_2O$$

(5) $$(2 H_2O + SO_2 \rightarrow SO_4^{2-} + 4 H^+ + 2 e^-) \times 4$$

(6) $$8 H_2O + 4 SO_2 \rightarrow 4 SO_4^{2-} + 16 H^+ + 8 e^-$$

$$8 e^- + 10 H^+ + 2 NO_3^- \rightarrow N_2O + 5 H_2O$$

$$\overline{8 H_2O + 4 SO_2 + 8 e^- + 10 H^+ + 2 NO_3^- \rightarrow 4 SO_4^{2-} + 16 H^+ + 8 e^- + N_2O + 5 H_2O}$$

In addition to the 8 e⁻, we can also cancel 5 H_2O and 10 H^+ from the equation leaving

$$3 H_2O(l) + 4 SO_2(g) + 2 NO_3^-(aq) \rightarrow 4 SO_4^{2-}(aq) + 6 H^+(aq) + N_2O(g)$$

Example 3.

The following reaction takes place in basic solution:

$$MnO_4^-(aq) + SO_3^{2-}(aq) \rightarrow MnO_2(s) + SO_4^{2-}(aq)$$

Write a complete, balanced equation for the reaction.

Solution.

(1) $$MnO_4^- \rightarrow MnO_2$$

$$SO_3^{2-} \rightarrow SO_4^{2-}$$

(2) Not needed

(3) $$4 H^+ + MnO_4^- \rightarrow MnO_2 + 2 H_2O$$

$$H_2O + SO_3^{2-} \rightarrow SO_4^{2-} + 2 H^+$$

(4) $$3 e^- + 4 H^+ + MnO_4^- \rightarrow MnO_2 + 2 H_2O$$

$$H_2O + SO_3^{2-} \rightarrow SO_4^{2-} + 2 H^+ + 2 e^-$$

Note that the charge isn't always zero on one side of a half-reaction. In the latter reaction, we add 2 e⁻'s to the right side to give a charge of 2– to both sides.

(5) $$(3 e^- + 4 H^+ + MnO_4^- \rightarrow MnO_2 + 2 H_2O) \times 2$$

$$(H_2O + SO_3^{2-} \rightarrow SO_4^{2-} + 2 H^+ + 2 e^-) \times 3$$

(6) $$6 e^- + 8 H^+ + 2 MnO_4^- \rightarrow 2 MnO_2 + 4 H_2O$$

$$\overline{3 H_2O + 3 SO_3^{2-} \rightarrow 3 SO_4^{2-} + 6 H^+ + 6 e^-}$$

$$6\,\bar{e} + 8\,H^+ + 2\,MnO_4^- + 3\,H_2O + 3\,SO_3^{2-} \rightarrow 2\,MnO_2 + 4\,H_2O + 3\,SO_4^{2-} + 6\,H^+ + 6\,e^-$$

$$2\,H^+ + 2\,MnO_4^- + 3\,SO_3^{2-} \rightarrow 2\,MnO_2 + H_2O + 3\,SO_4^{2-}$$

In basic solutions, an additional manipulation is needed. This may be performed after either step (3), step (4), or step (6). Any H^+ would be neutralized by OH^- in basic solutions. We can accomplish this by adding an equal number of OH^- to both sides and cancelling H_2O's. In this example:

$$2\,H^+ + 2\,MnO_4^- + 3\,SO_3^{2-} \rightleftarrows 2\,MnO_2 + H_2O + 3\,SO_4^{2-}$$

$$+\,2\,OH^- \qquad\qquad\qquad\qquad\qquad +\,2\,OH^-$$

$$\overline{2\,H_2O + 2\,MnO_4^- + 3\,SO_3^{2-} \rightleftarrows 2\,MnO_2 + H_2O + 3\,SO_4^{2-} + 2\,OH^-}$$

Cancelling out one H_2O, we are then left with the correct balanced equation:

$$H_2O\,(l) + 2\,MnO_4^-\,(aq) + 3\,SO_3^{2-}\,(aq) \rightleftarrows 2\,MnO_2\,(s) + 3\,SO_4^{2-}\,(aq) + 2\,OH^-\,(aq)$$

CATEGORY II   ELECTROCHEMICAL CELLS AND STANDARD VOLTAGES

There are two types of electrochemical cells. One type, called *electrolytic cells*, uses direct electric current to force non-spontaneous oxidation-reduction reactions to occur, i.e., they convert electrical energy into chemical energy. The second type, traditionally called *voltaic* (or galvanic) *cells*, uses spontaneous chemical oxidation-reduction reactions to produce a direct electric current, i.e., they convert chemical energy into electrical energy.

Let's consider the latter type of cell first. In a spontaneous oxidation-reduction reaction such as $Mg + 2\,H^+ \rightleftarrows Mg^{2+} + H_2$, there is a transfer of electrons from the substance being oxidized (e.g., Mg) to the substance being reduced (e.g., H^+). In a voltaic cell, the oxidation and reduction "half" reactions are separated and the electrons are forced to pass through an electrical circuit and perform some work before completing the chemical reaction. This is the principle behind the familiar "battery" that we all find so useful in our everyday lives.

Let's use the oxidation of Mg by H^+ as an example reaction and diagram a typical voltaic cell (see Fig. 12–1). In the left compartment there is a piece of metallic magnesium dipping into a solution containing Mg^{2+} ions (from MgX_2, where $X = Cl^-$, NO_3^-, etc.). In the right compartment H_2 gas is in contact with H^+ ions, (from HY, where $Y = Cl^-$, NO_3^-, etc.). The compartments are separated by a barrier that allows free passage of ions, but only slow and slight mixing of the solutions. This barrier is usually a salt bridge or a porcelain partition. The two compartments are connected by a wire, shown in the figure as connected through a voltmeter, V. When the wire is hooked up, the following occurs:

(a) Oxidation takes place at the Mg electrode: $Mg \rightarrow Mg^{2+} + 2\,e^-$. In the jargon of electrochemistry, the electrode at which oxidation occurs is called the *anode*.

(b) The electrons migrate through the wire and the voltmeter to the hydrogen electrode where reduction takes place: $2\,H^+ + 2\,e^- \rightarrow H_2$. The electrode at which reduction occurs is called the *cathode*.

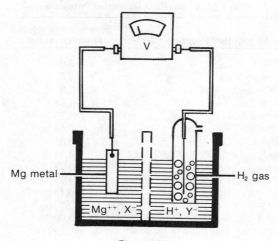

Figure 12–1

(c) To complete the circuit, and maintain electrical neutrality, negative Y^- ions pass through the barrier toward the anode and positive Mg^{2+} ions move through the barrier toward the cathode.

(d) The voltmeter measures the potential energy difference before and after electron transfer. This "potential difference" is called the voltage. Voltage is often called electromotive force, or emf, and is expressed in volts (1 volt = 1 joule/coulomb).

In our example cell, if $[Mg^{2+}] = [H^+] = 1.0$ M and the pressure of the H_2 (g) were 1.0 atm, the voltmeter would register 2.37 volts. If we substituted Zn metal and Zn^{2+} for the Mg and Mg^{2+}, the voltage of the cell would be 0.76 volt instead of 2.37 V. Using Sn and Sn^{2+} the voltage would be 0.14 V. If Cu metal and Cu^{2+} were used in the cell, the voltmeter would read 0.34 V, but would indicate that the flow of electrons was from the hydrogen to the copper, rather than from the metal to the hydrogen. Thus, with Cu and Cu^{2+} in the cell, the reaction is $Cu^{2+} + H_2 \rightleftarrows Cu + 2\ H^+$. Voltages are "differences," and obviously, different substances have different potentials for being oxidized or reduced. When only a difference can be measured it is convenient to relate quantities (energy, distance, time, etc.) to some arbitrary standard. Chemists have done this with electrode potentials. The *standard electrode potential* for the oxidation or reduction of hydrogen ($2\ H^+ + 2\ e^- \rightleftarrows H_2$) has been assigned a value of 0.00 volts. Other standard potentials may then be assigned relative to hydrogen. For example, since the potential difference in the Mg vs. H_2 cell is 2.37 volts, the standard oxidation potential for $Mg \rightarrow Mg^{2+} + 2\ e^-$ is assigned a value of +2.37 V. Magnesium has a positive tendency to be spontaneously oxidized relative to hydrogen. The standard reduction potential for $Mg^{2+} + 2\ e^- \rightarrow Mg$ is –2.37 V, and magnesium has a negative tendency to be spontaneously reduced compared to hydrogen. On the other hand, the standard oxidation potential for $Cu \rightleftarrows Cu^{2+} + 2\ e^-$ is –0.34 V and the standard reduction potential is +0.34 V. In this manner, an extensive table of standard potentials has been compiled. Table 12–1 gives a partial listing of reduction potentials. The values are based essentially on ion concentrations of 1.0 M, a temperature of 25°C, and gas pressures of 1 atm. Note that any standard oxidation potential (SOP) = – standard reduction potential (SRP).

Table 12–1   Standard Reduction Potentials

Reduction Half-Reaction	Standard Reduction Potential (volts)
$Li^{2+} + e^- \rightarrow Li$	-3.05
$Na^+ + e^- \rightarrow Na$	-2.71
$Mg^{2+} + 2\,e^- \rightarrow Mg$	-2.37
$Al^{3+} + 3\,e^- \rightarrow Al$	-1.66
$Zn^{2+} + 2\,e^- \rightarrow Zn$	-0.76
$Fe^{2+} + 2\,e^- \rightarrow Fe$	-0.44
$Cr^{3+} + e^- \rightarrow Cr^{2+}$	-0.41
$PbSO_4 + 2\,e^- \rightarrow Pb + SO_4{}^{2-}$	-0.36
$Co^{2+} + 2\,e^- \rightarrow Co$	-0.28
$Ni^{2+} + 2\,e^- \rightarrow Ni$	-0.25
$Sn^{2+} + 2\,e^- \rightarrow Sn$	-0.14
$Pb^{2+} + 2\,e^- \rightarrow Pb$	-0.13
$2\,H^+ + 2\,e^- \rightarrow H_2$	0.00
$Sn^{4+} + 2\,e^- \rightarrow Sn^{2+}$	$+0.15$
$Cu^{2+} + e^- \rightarrow Cu^+$	$+0.15$
$Cu^{2+} + 2\,e^- \rightarrow Cu$	$+0.34$
$Fe^{3+} + e^- \rightarrow Fe^{2+}$	$+0.77$
$Ag^+ + e^- \rightarrow Ag$	$+0.80$
$Br_2 + 2\,e^- \rightarrow 2\,Br^-$	$+1.07$
$O_2 + 4\,H^+ + 4\,e^- \rightarrow 2\,H_2O$	$+1.23$
$Au^{3+} + 3\,e^- \rightarrow Au$	$+1.50$
$F_2 + 2\,e^- \rightarrow 2\,F^-$	$+2.87$

Standard potentials are used to predict which oxidation-reduction reactions will occur spontaneously and what voltages will be produced from voltaic cells. Any spontaneous reaction must have a positive voltage.

Example 1.

Write the reactions which would occur and give the standard voltages produced if the following metals and metal ions were used in place of Mg and Mg^{2+} in the cell shown in Figure 12–1: (a) Al and Al^{3+} (b) Ag and Ag^+ (c) Ni and Ni^{2+}.

Solution.

(a) Aluminum has a negative reduction potential relative to hydrogen, so hydrogen would be reduced in the cell. The reaction would be $2\,Al + 6\,H^+ \rightleftarrows 2\,Al^{3+} + 3\,H_2$, and the voltage produced (for $[Al^{3+}] = 1.0$ M) would be 1.66 volts.

(b) Silver has a positive reduction potential of 0.80 V. The cell reaction would be $H_2 + 2\,Ag^+ \rightleftarrows 2\,H^+ + Ag$, and the cell voltage would be 0.80 V.

(c) The standard reduction potential for nickel is -0.25 V, so the reaction would be $Ni + 2\,H^+ \rightleftarrows Ni^{2+} + H_2$, and the cell voltage would be 0.25 V.

Example 2.

A voltaic cell is constructed using Al and Al^{3+} in one half-cell, and Ag and Ag^+ in the other half-cell. (a) What total reaction will occur? (b) What half-reaction will occur at each electrode? (c) Which is the anode and which is the cathode? (d) How many volts will the cell produce (if $[Al^{3+}]$ and $[Ag^+]$ are 1.0 M)?

Solution.

(a) The reduction potential for Ag is positive relative to hydrogen, and the reduction potential for Al is negative. Therefore, silver (Ag^+ ions) will be reduced and Al will be oxidized. The balanced equation for the reaction is

$$Al + 3\,Ag^+ \rightleftharpoons Al^{3+} + 3\,Ag$$

(b) The half-reactions are

$$Al \rightarrow Al^{3+} + 3\,e^-$$

$$3\,Ag^+ + 3\,e^- \rightarrow 3\,Ag$$

(c) Since oxidation occurs at the Al electrode, that is called the anode. Reduction occurs at the Ag electrode, and that is the cathode.

(d) The voltage, or potential difference, produced by the cell is just the total difference between the standard potentials for Al and Ag. The standard reduction potential for Al is -1.66 V and for Ag it is $+0.80$. The total *difference* is 2.46 V, and this is the voltage produced by the cell.

Note two things: The (standard) potential difference produced by the cell does not depend upon the balanced equation for the cell reaction. Even though we need $3\,Ag^+ + 3\,e^- \rightarrow 3\,Ag$ to balance the equation for the reaction, this in no way changes the standard reduction potential for Ag. It is still $+0.80$ V, *not* $3 \times +0.80$ V. Standard potentials are relative to hydrogen, and the potential for reduction of H^+ to H_2 is 0.00 V, regardless of whether we write the half-reaction as $2\,H^+ + 2\,e^- \rightarrow H^2$, or $H^+ + e^- \rightarrow \frac{1}{2}\,H$, or $4\,H^+ + 4\,e^- \rightarrow 2\,H_2$. Also, notice that the total cell voltage may be conveniently obtained by simply adding the standard potentials for the half-reactions as they actually occur in the cell. In this cell, the Al is oxidized. The standard potential for the reduction of Al^{3+} to Al is -1.66 V; therefore, as an oxidation the standard potential is $+1.66$ V. In short, SOP $= -$SRP.

Thus, writing the half-reactions as they occur, the total cell voltage is just the sum of the standard potentials for the half reactions:

Reaction	*Standard Potential*
$Al \rightarrow Al^{3+} + 3\,e^-$	$+1.66$ V
$3\,Ag^+ + 3\,e^- \rightarrow 3\,Ag$	$+0.80$ V
$Al + 3\,Ag^+ \rightarrow Al^{3+} + 3\,Ag$	$+2.46$ V

By convention, any spontaneous oxidation-reduction reaction will have a positive standard potential (i.e., will produce a positive cell voltage).

Example 3.

A cell is constructed using Zn and Zn^{2+} in one half-cell and Ni with Ni^{2+} in the other half-cell. Write the reaction and find the standard potential of the cell.

Solution.

Both reduction potentials are negative, but the zinc potential is more negative. Therefore, the half-reactions, the total reaction, and the standard potentials are

$Ni^{2+} + 2\,e^- \rightarrow Ni$	−0.25 V
$Zn \rightarrow Zn^{2+} + 2\,e^-$	+0.76 V
$Ni^{2+} + Zn \rightarrow Ni + Zn^{2+}$	+0.51 V

The nickel (NI^{2+}) has a *more* positive tendency to be reduced, so the nickel is reduced and the zinc is oxidized.

The other type of electrochemical cell is the electrolytic cell. In an electrolytic cell an electric current is used to produce a chemical reaction, instead of vice-versa. Figure 12–2 is a simple schematic of an electrolytic cell.

The direct current source acts as an electron pump, supplying electrons at the cathode and removing them from the anode. The electrodes are usually some conducting, but chemically inert, material such as graphite or platinum. The electrolyte may be any conducting liquid in which one wishes to force a non-spontaneous oxidation-reduction reaction to occur. For example, the electrolyte could be molten NaCl. If that were so, the "forced" reactions would be: reduction at the cathode, $Na^+ + e^- \rightarrow Na(s)$, and oxidation at the anode, $Cl^- \rightarrow \frac{1}{2}\,Cl_2\,(g) + e^-$. The net reaction occurring in the cell would be

$$Na^+ + Cl^- \rightarrow Na(s) + \tfrac{1}{2}\,Cl_2\,(g)$$

In fact, thousands of tons of sodium metal and chlorine gas are produced commercially by electrolysis of molten sodium chloride. How else could you produce metallic sodium?

Electrolysis reactions may be carried out in aqueous solutions, but, of course, water may also be "broken down" electrolytically into H_2 and O_2, and competitive oxidation-reduction reactions can occur in aqueous solutions. For example, electrolysis of aqueous NaCl produces H_2 at the cathode, not Na, because the hydrogen (oxidation state +1) of water molecules is more easily reduced than aqueous Na^+ ions.

Reactions occurring in electrolytic cells may be described quantitatively by *Faraday's Laws.* If one multiplies the charge per electron (1.602×10^{-19}

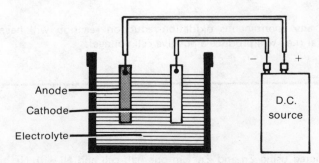

Figure 12–2

coulomb) times the number of electrons per mole (6.023×10^{23}), a quantity called the faraday is obtained.

$$1 \text{ faraday} = 1 \text{ mole of } e^-\text{'s} = 96,500 \text{ coulombs}$$

The ampere is a measure of rate of flow of electricity. If one coulomb passes a point in a circuit in one second, then there is an electrical "current" of one ampere, i.e.,

$$\text{no. of coulombs} = (\text{no. of amperes}) \times (\text{no. of seconds})$$

With these two laws it is possible to predict and describe the stoichiometry of electrolytic cells.

Example 4.

How many grams of Na and Cl_2 are produced when a current of 1.0 ampere is passed through molten NaCl for 1.0 hour?

Solution.

First, find the number of coulombs (amps $\times$ sec).

$$X \text{ sec} = 1.0 \text{ hr} \times \frac{60 \text{ min}}{1 \text{ hr}} \times \frac{60 \text{ sec}}{1 \text{ min}} = 3600 \text{ sec}$$

and,

$$\text{no. of coulombs} = 1.0 \text{ amp} \times 3600 \text{ sec} = 3600 \text{ coulombs}$$

Then, knowing the half-reactions at each electrode:

$$Na^+ + e^- \rightarrow Na$$

$$Cl^- \rightarrow \tfrac{1}{2} Cl_2 + e^-$$

the problem may be conveniently solved by the factor-label method.

$$X \text{ g Na} = 3600 \text{ coulombs} \times \frac{1 \text{ mole of } e^-\text{'s}}{96,500 \text{ coulombs}} \times \frac{1 \text{ mole Na}}{1 \text{ mole of } e^-\text{'s}} \times$$

$$\frac{23.0 \text{ g Na}}{1 \text{ mole Na}} = 0.858 \text{ gram Na}$$

$$X \text{ g Cl}_2 = 3600 \text{ coulombs} \times \frac{1 \text{ mole of } e^-\text{'s}}{96,500 \text{ coulombs}} \times \frac{\tfrac{1}{2} \text{ mole of Cl}_2}{1 \text{ mole of } e^-\text{'s}} \times \frac{71.0 \text{ g Cl}_2}{1 \text{ mole Cl}_2}$$

$$= 1.32 \text{ grams Cl}_2$$

Thus, quantitative electrolysis problems are similar to other types of stoichiometry problems, with "moles of electrons" appearing as either reactants or products. Note the second conversion factor in the factor-label solution.

Example 5.

A piece of jewelry is to be "gold plated" by electrolytic reduction of Au from a

$Au(NO_3)_3$ solution. (In electroplating, the object to be plated is used as the cathode.) At a current of 0.50 amp, how much time is required to plate out 0.30 gram of gold onto the object?

Solution.

The cathode reaction will be

$$Au^{3+} + 3\,e^- \rightarrow Au$$

and we can work "backwards" from the approach used in the previous example.

$$X \text{ coulombs} = 0.30 \text{ g Au} \times \frac{1 \text{ mole Au}}{197.0 \text{ g Au}} \times \frac{3 \text{ moles of } e^-\text{'s}}{1 \text{ mole Au}} \times \frac{96{,}500 \text{ coulombs}}{1 \text{ mole of } e^-\text{'s}}$$

$$= 440.9 \text{ coulombs}$$

$$\text{no. of coulombs} = \text{amps} \times \text{sec}$$

$$\text{sec} = 440.9 \text{ coulombs} \div 0.50 \text{ amp} = 882 \text{ sec or}$$

$$882 \text{ sec} \div 60 = 14.7 \text{ minutes}$$

Example 6.

The gram equivalent weight (GEW) of a substance in an oxidation-reduction reaction is the weight of the substance that will lose or gain one mole of electrons. In an electrolytic cell 2.34 grams of an unknown metal is deposited at the cathode by a current of 3.50 amps operating for 18.0 minutes. Find the gram equivalent weight of the metal. If the cathode reaction is $M^{2+} + 2\,e^- \rightarrow M$, what is the atomic weight of the metal?

Solution.

$$X \text{ sec} = 18.0 \text{ min} \times \frac{60 \text{ sec}}{1 \text{ min}} = 1080 \text{ sec}$$

$$\text{no. of coulombs} = 3.50 \text{ amps} \times 1080 \text{ sec} = 3780 \text{ coulombs}$$

The GEW of the metal will be the weight of metal that is equivalent to one mole of electrons, so we can solve:

$$X \text{ g M} = 1 \text{ mole of } e^-\text{'s} \times \frac{96{,}500 \text{ coulombs}}{1 \text{ mole of } e^-\text{'s}} \times \frac{2.34 \text{ g M}}{3780 \text{ coulombs}} = 59.7 \text{ g M}$$

If the cathode reaction were $M^{2+} + 2\,e^- \rightarrow M$, then we see that one mole of electrons would deposit ½ mole of M metal, and the weight of one mole of M must be $2 \times 59.7 = 119.4$ g (the atomic weight).

CATEGORY III   SPONTANEITY AND EXTENT OF REDOX REACTIONS

Oxidation-reduction reactions are spontaneous or they are not. Thermodynamics tells us that spontaneity is a function of free energy, ΔG. It follows that

ΔG and the standard potential for an oxidation-reduction reaction are related. The relationship is

$$\Delta G^\circ = - n F E^\circ$$

where n is the number of electrons transferred in the reaction, F is the faraday (96,500 coulombs), and E° is the standard potential for the reaction. For a direct conversion between ΔG in kcal and E° in volts, the equation is

$$\Delta G^\circ = -23.06 \, n \, E^\circ \tag{1}$$

Standard potentials can also be related to equilibrium constants. The relationship is

$$\Delta G^\circ = -23.06 \, n \, E^\circ = - RT \ln K$$

or,

$$n \, E^\circ = .0592 \log K \tag{2}$$

for a temperature of 25°C and using base 10 for logarithms.

Equations (1) and (2) will be used for quantitative problems concerning spontaneity and extent of oxidation-reduction reactions, but let's first examine qualitative predictions of spontaneity of redox reactions. This is an area of real concern to an experimental chemist. What will react with what (spontaneously)? Will Pb reduce Ag^+ ions to silver metal? Will Cr metal be oxidized by water? The chemist can find the answers to such questions by referring to a table of standard potentials. If reduction potentials are used, we follow what we can call an "AC rule" (AC for anticlockwise). Given a list of reduction potentials (Table 12–1), spontaneous reactions will occur only according to the anti-clockwise arrow shown in Figure 12–3. Li will be oxidized by F_2 to give Li^+ and F_2 will be reduced to give F^-. Examine Table 12–1. Mg will react (spontaneously) with H^+ to give Mg^{2+} and H_2; Zn will react with Ni^{2+} to give Zn^{2+} and Ni, etc. Any "reducing agent" higher and on the right-hand side of the table will react with any "oxidizing agent" lower and on the left-hand side of the table. Under standard conditions, all oxidation-reduction reactions will obey the AC rule. It is a qualitative device to help us remember the fundamental point that a redox reaction will be spontaneous if the sum of the appropriate potentials is positive.

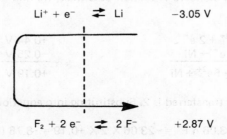

Figure 12–3

Example 1.

Which of the following reactions will occur spontaneously? Refer to Table 12–1 of reduction potentials.

(a) $Fe + Co^{2+} \rightleftarrows Fe^{2+} + Co$

(b) $Ni + Sn^{2+} \rightleftarrows Ni^{2+} + Sn$

(c) $Au + 3\,Ag^+ \rightleftarrows Au^{3+} + 3\,Ag$

(d) $Br_2 + Cu^{2+} \rightleftarrows 2\,Br^- + Cu$

(e) $2\,Cu^+ \rightleftarrows Cu + Cu^{2+}$

(f) $2\,Br^- + \frac{1}{2}\,O_2 + 2\,H^+ \rightleftarrows Br_2 + H_2O$

(g) $H_2 + Sn^{2+} \rightleftarrows 2\,H^+ + Sn^{4+}$

Solution.

(a) Yes. Following the AC rule, Fe would be oxidized to Fe^{2+} while Co^{2+} is reduced to Co.

(b) Yes. Again, the reaction obeys the AC rule.

(c) No. While Au is oxidized and Ag^+ reduced, the AC rule tells us that this reaction will not occur *spontaneously.*

(d) No! Two things cannot be reduced simultaneously.

(e) Yes. The AC rule shows that $2\,Cu^+$ can react to give Cu and Cu^{2+}. This type of reaction, where a substance reacts with itself, is called "disproportionation."

(f) Yes. The AC rule predicts that Br^-, in an acid solution, will react with O_2 to give Br_2 and H_2O.

(g) No way! Two things cannot be simultaneously oxidized. Remember to follow that AC arrow all the way around the table.

Now let's examine the quantitative use of equations (1) and (2).

$$\Delta G° = -23.06\,n\,E° \qquad (1)$$

$$nE° = .0592 \log K \qquad (2)$$

Example 2.

Find $\Delta G°$ and the equilibrium constant, K, for the oxidation-reduction reaction

$$Fe + Ni^{2+} \rightleftarrows Fe^{2+} + Ni$$

Solution.

Referring to the table of standard reduction potentials, we find $E°$ for the reaction to be

$Fe \rightarrow Fe^{2+} + 2\,e^-$	+0.44 V
$Ni^{2+} + 2\,e^- \rightarrow Ni$	−0.25 V
$Fe + Ni^{2+} \rightleftarrows Fe^{2+} + Ni$	+0.19 V

The number of electrons transferred is 2. Substituting into equation (1), we find,

$$\Delta G° = -23.06\,n\,E° = -23.06 \times 2 \times +0.19 = -8.76 \text{ (kcal)}$$

Substitution into equation (2) enables us to calculate the equilibrium constant, K, for the reaction

$$n \, E° = .0592 \log K; \quad \log K = \frac{2 \times +0.19}{.0592}; \quad \log K = 6.42$$

$$K = 10^6 \times \text{antilog of } .42, \text{ or } K = 2.6 \times 10^6$$

$\Delta G°$, $E°$, and K all tell us that the reaction would occur spontaneously.

Example 3.

A cell is constructed that uses the following half reactions:

$$Au(s) + 4 \, Cl^- \rightarrow AuCl_4^- + 3 \, e^-$$

$$Au^{3+} + 3 \, e^- \rightarrow Au(s)$$

The standard potential of the cell is found to be +0.50 volt. Find the equilibrium constant for the reaction

$$Au^{3+} + 4 \, Cl^- \rightleftarrows AuCl_4^-$$

Solution.

Often cell potentials afford a very useful and relatively simple means of determining equilibrium constants. In this example, the standard potential for the oxidation is −1.00 V, for the reduction it is +1.50 V, and $E°$ for the cell is +0.50 V.

$Au(s) + 4 \, Cl^- \rightarrow AuCl_4^- + 3 \, e^-$	− 1.00 V
$Au^{3+} + 3 \, e^- \rightarrow Au(s)$	+1.50 V
$Au^{3+} + 4 \, Cl^- \rightarrow AuCl_4^-$	+0.50 V

The number of electrons transferred in the cell is 3. Substitution into equation (2) gives

$$n \, E° = .0592 \log K; \quad \log K = \frac{(3) \times (+0.50)}{.0592} = 25.34$$

$$\log K = 25 + .34; \quad K = 2.2 \times 10^{25}$$

The equilibrium lies far to the right and $AuCl_4^-$ is a very stable complex ion.

Consider an oxidation-reduction equilibrium, such as the reaction in Example 2, $Fe + Ni^{2+} \rightleftarrows Fe^{2+} + Ni$. What would happen to the equilibrium if we increased the concentration of Ni^{2+}? Equilibrium principles tell us that the reaction would "shift" to the right, i.e., there would be an increased driving force for the reaction to proceed from left to right. Or, if we increased (conc. Fe^{2+}), the equilibrium would shift to the left, and so on. It is not surprising that changes in concentration of reactants and/or products will change the voltage produced by an oxidation-reduction reaction in a voltaic cell. Increasing the concentration of reactants (or decreasing the concentration of products) provides an increased driving force for reaction and therefore increases the potential. Increasing the concentration of products, or decreasing the concentration of reactants, decreases the potential. This effect is described by the *Nernst equation*:

$$E = E° - \frac{.0592}{n} \log Q$$

where E is the potential at concentrations other than 1.0 M, E° is the standard potential, and Q is the equilibrium quotient. Q is identical in form to the equilibrium constant expression, but the concentrations used in Q are *not* equilibrium concentrations.

Example 4.

A voltaic cell is constructed to utilize the reaction $Fe + Ni^{2+} \rightleftarrows Fe^{2+} + Ni$. In the cathode half-cell (conc. Ni^{2+}) is 2.0 M and in the anode half-cell (conc. Fe^{2+}) is 0.50 M. Find the voltage produced by the cell.

Solution.

In Example 2 we found that the standard potential, E°, for the reaction is +0.19 V. The number of electrons transferred, n, is 2. The equilibrium quotient, Q, is

$$\frac{(conc. Fe^{2+})(conc. Ni)}{(conc. Fe)(conc. Ni^{2+})} = \frac{(conc. Fe^{2+})}{(conc. Ni^{2+})}$$

since the concentrations of solids (metallic Fe and Ni) do not appear in the expression. We can now substitute into the Nernst equation:

$$E = E° - \frac{.0592}{n} \log \frac{(conc. Fe^{2+})}{(conc. Ni^{2+})}$$

$$= +0.19 V - \frac{.0592}{2} \log \frac{(0.50)}{(2.0)}$$

$$= +0.19 V - (.0296) \log 0.25 = +0.19 V - (-0.018) = +0.21 V$$

The log of $2.5 \times 10^{-1} = -1 + \log 2.5 = -1 + .398 = -0.602$. Notice that even fairly drastic changes in concentrations do not affect the cell potential a great deal. The standard potential is +0.19 V, and by increasing (conc. Ni^{2+}) to 2.0 M and decreasing (conc. Fe^{2+}) to 0.50 M, we only increase the potential by 0.02 V.

Example 5.

A voltaic cell is set up, using Ag and Pb electrodes, to use the reaction

$$Pb + 2 Ag^+ \rightleftarrows Pb^{2+} + 2 Ag$$

In the lead half-cell, the concentration of Pb^{2+} is known to be 1.00 M. The concentration of Ag^+ in the silver half-cell is not known. The cell registers a voltage of +0.89 volt. Find (conc. Ag^+).

Solution.

From the table of reduction potentials we find that the standard potential for the cell is +0.93 V. From the Nernst equation we know that

$$E = E° - \frac{.0592}{n} \log \frac{(\text{conc. } Pb^{2+})}{(\text{conc. } Ag^+)^2}$$

The number of electrons transferred is 2, and substituting, we find

$$+ 0.89 \text{ V} = +0.93 \text{ V} - (.0296) \log \frac{(1.00)}{(\text{conc. } Ag^+)^2}$$

$$\log \frac{(1.00)}{(\text{conc. } Ag^+)^2} = \log 1.00 - \log (\text{conc. } Ag^+)^2 = \frac{-0.04}{-.0296} = 1.35$$

$$\log (\text{conc. } Ag^+)^2 = -1.35 = (0.65 - 2); (\text{conc. } Ag^+)^2 = 4.47 \times 10^{-2}$$

$$(\text{conc. } Ag^+) = 0.21 \text{ M}$$

Thus, we see that cell potential affords a sensitive means of measuring ionic concentration in solution. This is the principle behind valuable laboratory tools, such as the pH meter.

Category I   **PROBLEMS**

1. Complete and balance the following oxidation-reduction reactions (in acid solution).
    (a) $I^- + NO_3^- \rightleftarrows I_2 + NO$
    (b) $SO_2 + Cr_2O_7^{2-} \rightleftarrows SO_4^{2-} + Cr^{3+}$
    (c) $ClO_3^- \rightleftarrows ClO_4^- + Cl^-$
    (d) $F_2 + H_2O \rightleftarrows F^- + O_2$
    (e) $S_2O_3^{2-} + MnO_4^- \rightleftarrows S_4O_6^{2-} + Mn^{2+}$
    (f) $C_2H_5OH + BrO_3^- \rightleftarrows CH_3COOH + Br^-$
    (g) $CH_3CHO + O_2 \rightleftarrows CO_2 + H_2O$

2. Complete and balance the following reactions (in basic solution).
    (a) $Br^- + CrO_4^{2-} \rightleftarrows BrO_3^- + Cr_2O_3$
    (b) $ClO^- + SO_3^{2-} \rightleftarrows Cl^- + SO_4^{2-}$
    (c) $C_2O_4^{2-} + MnO_4^{2-} \rightleftarrows CO_3^{2-} + MnO_2$
    (d) $CrI_3 + Cl_2 \rightleftarrows CrO_4^{2-} + IO_4^- + Cl^-$
    (e) $SnO + CrO_4^{2-} \rightleftarrows Sn(OH)_6^{2-} + Cr(OH)_4^-$
    (f) $C_2H_5OH + BH_4^- \rightleftarrows C_2H_6 + BO_2^-$
    (g) $ClO_3^- \rightleftarrows ClO_4^- + Cl^-$

3. If an unknown acidic Fe^{2+} solution was "titrated" with a 0.50 M $Cr_2O_7^{2-}$ solution until all of the Fe^{2+} was converted to Fe^{3+}, and the titration required 46.0 ml of the $Cr_2O_7^{2-}$ solution, how many moles of Fe^{2+} were present in the unknown solution? ($Cr_2O_7^{2-} \rightarrow Cr^{3+}$).

Category II

4. A voltaic cell uses Co and Co^{2+} in one half-cell and Au with Au^{3+} in the other half-cell. (a) Sketch a diagram of the cell; (b) write the half-reactions occurring at the anode and cathode; (c) determine the voltage produced at standard conditions; (d) identify the direction of electron flow, and movement of positive and negative ions.

5. In the lead storage cell (car battery), the complete reaction is $Pb + PbO_2 + 4\ H^+ + 2\ SO_4^{2-} \rightleftarrows 2\ PbSO_4 + 2\ H_2O$. Each cell produces 2.00 volts. Determine the reaction at the cathode and the standard reduction potential for the cathode reaction, assuming standard conditions.

6. Based upon the standard potentials listed in this chapter, what type of voltaic cell would produce the maximum possible voltage? How would you design such a cell?

7. A voltaic cell consists of Fe and Fe^{2+} in one half-cell, and Sn with Sn^{2+} in the other. (a) Write the complete cell reaction; (b) determine which is the anode and which is the cathode; (c) calculate the voltage of the cell.

8. Electrolysis of aqueous silver nitrate, $AgNO_3$, produces Ag metal at the cathode and O_2 gas at the anode. (a) Why? (b) How many grams of Ag would be produced from a current of 0.75 amp operating for 35 minutes?

9. The Hall process for production of Al metal uses electrolytic reduction of Al^{3+} from an Al_2O_3–Na_3AlF_6 melt. How many kilograms of aluminum could be produced using a current of 20,000 amps for 8 hours?

10. An electrolytic cell is used to "galvanize" iron. A thin coating of zinc, deposited on the surface of the Fe, protects it from air and moisture oxidation. Using a cell operating at 20.0 amps, how long would it take to deposit 327.0 grams of Zn on a piece of iron pipe? The electrolyte is a $ZnSO_4$ solution.

11. A solution containing Tc^{n+} was electrolyzed with a current of 5.0 amps. In 16.1 minutes, 2.50 grams of Tc metal were deposited. (a) What was the gram equivalent weight of the Tc^{n+}; (b) what was the oxidation state of the Tc^{n+} in the solution?

Category III

12. From the table of reduction potentials, predict which of the following reactions should occur spontaneously:
    (a) $Sn + Sn^{4+} \rightleftarrows 2\ Sn^{2+}$
    (b) $4\ Cr^{2+} + Sn^{4+} \rightleftarrows Sn + 4\ Cr^{3+}$
    (c) $H_2 + 2\ Fe^{2+} \rightleftarrows 2\ H^+ + 2\ Fe^{3+}$
    (d) $Pb^{2+} + Sn^{2+} \rightleftarrows Pb + Sn^{4+}$
    (e) $Cu^+ + Ag^+ \rightleftarrows Cu^{2+} + Ag$
    (f) $F_2 + Mg^{2+} \rightleftarrows 2\ F^- + Mg$
    (g) $Pb + 2\ H^+ \rightleftarrows Pb^{2+} + H_2$

13. Calculate $\Delta G°$ and the equilibrium constant for the disproportionation reaction

$$2\ Cu^+ \rightleftarrows Cu + Cu^{2+}$$

Be careful! How many electrons are transferred in the reaction?

14. Find the equilibrium constant for the reaction $Fe^{2+} + Ag^+ \rightleftarrows Fe^{3+} + Ag$.

15. Cobalt and nickel are notoriously similar in most of their chemistry. Find the equilibrium concentration of Ni^{2+} if 1.0 mole Ni^{2+} is added to a 1.0 liter solution originally 2.00 M in Co^{2+}. (Ni and Co also present.)

16. Calculate K and $\Delta G°$ for the reaction

$$2\ Cr^{2+} + Fe^{2+} \rightleftarrows 2\ Cr^{3+} + Fe$$

17. A voltaic cell utilizes the reaction

$$Zn + Sn^{2+} \rightleftarrows Zn^{2+} + Sn$$

What is the potential of the cell when
(a) (conc. Sn^{2+}) = 1.0 M; (conc. Zn^{2+}) = 1.0 M?
(b) (conc. Sn^{2+}) = 3.0 M; (conc. Zn^{2+}) = 0.15 M?
(c) (conc. Sn^{2+}) = 0.20 M; (conc. Zn^{2+}) = 2.0 M?

18. A voltaic cell is set up, under standard conditions, using the reaction

$$Pb + Cu^{2+} \rightleftarrows Pb^{2+} + Cu$$

Then Cl^- is added to the lead half-cell until voltage increases to +0.55 V. As Cl^- is added, $PbCl_2$ precipitates. Find the K_{sp} of $PbCl_2$. The equilibrium concentration of Cl^- is 0.09 M.

General Problems

19. A steady current is passed through a $CoSO_4$ solution until 2.35 g of metallic cobalt are produced. How many coulombs are used?

20. Which of the following will be spontaneously reduced by Pb?
(a) Co^{2+}
(b) Cu^{2+}
(c) H^+
(d) Br_2
(e) Zn^{2+}

21. Find the equilibrium constants for the following reactions:
(a) $Sn^{2+} + 2 H^+ \rightleftarrows Sn^{4+} + H_2$
(b) $Cr^{2+} + Fe^{3+} \rightleftarrows Cr^{3+} + Fe^{2+}$

22. Give the oxidation number (or oxidation state) of each atom in
(a) SO_3
(b) HIO_4
(c) $S_2O_3{}^{2-}$
(d) $H_4P_2O_7$
(e) $HBrO$
(f) NO^+

23. The reaction $M(s) + 3 N^+(aq) \rightleftarrows M^{3+}(aq) + 3 N(s)$ is at equilibrium when $[N^+]$ = 0.01 M and $[M^{3+}]$ = 0.004 M. Calculate K, $E°$, and $\Delta G°$ for the reaction.

24. A voltaic cell utilizes the reaction $Ag^+(aq) + e^- \rightarrow Ag(s)$ at one electrode and $Tl(s) \rightarrow Tl^+(aq) + e^-$ at the other. Under standard conditions the cell generates 1.14 volts. What is the standard reduction potential for the thallium half-reaction?

13 NUCLEAR REACTIONS

If you could go back in time and give a caveman a modern wristwatch and ask him to explain how it works, how would he solve the problem? The results would probably be similar to the attempts of twentieth century man to understand the atomic nucleus.

Since the dawn of atomic theory, man has been trying to understand what makes the nucleus "tick." It seems likely that the caveman would have approached his problem by striking the watch with a "high energy particle," such as a rock, and observing the fragments in hopes that the pieces would reveal the secret of the watch's operation. Modern man bombards the tiny nucleus with a variety of "rocks" (high velocity protons, deuterons, alpha particles, etc.), observes the fragments, and tries to understand how the different "pieces" relate to nuclear theory. To date, over 30 different nuclear particles, or fragments, have been detected, and simply trying to understand the nature of the various particles has proven as difficult for twentieth century man as understanding the nature of wheels, gears, dials, and batteries would have been for the caveman. One wonders what kind of theory the caveman would have developed to explain his observations.

We shall leave nuclear theory to the realm of nuclear physics and limit our observations to three categories of problems.

Problem Categories

I   Types of nuclear reactions.
II Rates of nuclear reactions.
III Mass-energy relations.

CATEGORY I   TYPES OF NUCLEAR REACTIONS

As far as today's knowledge extends, chemical reactions and nuclear reactions are independent of each other. Chemical reactions involve electrons around the nucleus. With few exceptions, nuclear reactions involve the emission of energy and one or more particles from the nucleus. Nuclear "decay" occurs whenever there is an unstable neutron/proton ratio in a nucleus. The most familiar types of nuclear emission (or "radiation") are

Alpha, α, radiation. Alpha particles are helium nuclei, consisting of two protons and two neutrons.

Beta, β^-, radiation. Beta (β^-) particles have the mass and charge of electrons and may be considered electrons that come out of the nucleus. Emission of a β^- particle from the nucleus converts a neutron into a proton.

Beta, β^+, radiation. Often called positrons, β^+ particles have the mass of electrons but carry a positive charge. Emission of a positron from the nucleus converts a proton into a neutron.

Gamma, γ, radiation. Gamma rays are high-energy photons. Emission of γ-rays change the energy, but not the particle make-up, of the nucleus. Virtually all nuclear reactions are accompanied by gamma radiation.

Neutron radiation. Simple neutron emission by a nucleus rarely occurs. However, neutrons are frequently used as bombarding particles to initiate nuclear reactions.

In writing nuclear reactions we will use the following symbols: Alpha particles = ^4_2He; Beta (β^-) particles = $^0_{-1}\text{e}$; positrons (β^+ particles) = ^0_1e; neutrons = ^0_1n.

Example 1.

Write balanced nuclear equations for
- (a)   alpha emission by $^{202}_{85}\text{At}$ followed by beta emission by the product
- (b)   positron emission by $^{195}_{79}\text{Au}$
- (c)   gamma emission by $^{122}_{54}\text{Xe}$

Solution.

Like chemical reactions, nuclear reactions must balance to yield equations. Note that in writing the symbols for various isotopes of elements we use the number of protons in the nucleus as the subscript (the atomic number) and the total number of protons and neutrons as the superscript (the mass number).

(a) $^{202}_{85}\text{At} \rightarrow {}^{198}_{83}\text{Bi} + {}^4_2\text{He}$

      $^{198}_{83}\text{Bi} \rightarrow {}^{198}_{84}\text{Po} + {}^0_{-1}\text{e}$

Loss of an alpha particle decreases the number of protons by two and the total number of protons and neutrons by four. Thus At is converted into an isotope of Bi. Loss of a beta particle converts a neutron into a proton. So, while the sum of neutrons plus protons remains constant, the number of protons increases by one, and in this example, an isotope of Bi becomes an isotope of Po. Mass numbers and atomic numbers must balance in nuclear reactions.

(b) $^{195}_{79}\text{Au} \rightarrow {}^{195}_{78}\text{Pt} + {}^0_1\text{e}$

Emission of a positron converts a proton into a neutron, hence the number of protons in the nucleus decreases by one while the mass number remains constant.

(c) $^{122}_{54}\text{Xe} \rightarrow {}^{122}_{54}\text{Xe} + \gamma(\text{energy})$

Gamma emission changes neither the atomic number nor the mass number. It only decreases the energy of the nucleus, making it (somewhat) more stable.

Thousands of fascinating nuclear reactions have been "induced" by man with his twentieth century nuclear "rocks." The first such induced process was the bombardment of nitrogen with alpha particles by Rutherford in 1919. He observed the reaction

$$^{14}_{7}N + ^{4}_{2}He \rightarrow ^{17}_{8}O + ^{1}_{1}H$$

Madam Curie studied several similar nuclear reactions, such as

$$^{27}_{13}Al + ^{4}_{2}He \rightarrow ^{30}_{15}P + ^{1}_{0}n$$

and helped establish the fact that man could, artificially, transform one element into another.

Example 2.

If we label the element to be bombarded as the "target" and identify the bombarding particle and the expelled (or product) particle, write balanced nuclear equations for the following reactions,

	Target	Bombarding Particle	Expelled Particle(s)
(a)	$^{40}_{18}Ar$	$^{4}_{2}He$	$^{1}_{1}H$
(b)	$^{130}_{52}Te$	$^{2}_{1}H$	$^{1}_{0}n$
(c)	$^{24}_{12}Mg$	$^{2}_{1}H$	$^{4}_{2}He$
(d)	$^{96}_{40}Zr$	$^{4}_{2}He$	$^{1}_{0}n$
(e)	$^{197}_{79}Au$	$^{2}_{1}H$	$2\,^{1}_{0}n$
(f)	$^{63}_{29}Cu$	$^{1}_{1}H$	$^{1}_{1}H, ^{1}_{0}n$

Solution

(a) $^{40}_{18}Ar + ^{4}_{2}He \rightarrow ^{43}_{19}K + ^{1}_{1}H$

Note that $^{1}_{1}H$ is just a proton. The symbol $^{2}_{1}H$ representa deuterium nucleus, one proton plus one neutron. Deuterium nuclei are often called deuterons.

(b) $^{130}_{52}Te + ^{2}_{1}H \rightarrow ^{131}_{53}I + ^{1}_{0}n$

(c) $^{24}_{12}Mg + ^{2}_{1}H \rightarrow ^{22}_{11}Na + ^{4}_{2}He$

(d) $^{96}_{40}Zr + ^{4}_{2}He \rightarrow ^{99}_{42}Mo + ^{1}_{0}n$

(e) $^{197}_{79}Au + ^{2}_{1}H \rightarrow ^{197}_{80}Hg + 2\,^{1}_{0}n$

(f) $^{63}_{29}Cu + ^{1}_{1}H \rightarrow ^{62}_{29}Cu + ^{1}_{1}H + ^{1}_{0}n$

Example 3.

Bombardment of $^{52}_{24}Cr$ with high-energy alpha particles yields two products, A and

B, according to the equation

$$2\, {}^{52}_{24}Cr + 2\, {}^{4}_{2}He \rightarrow A + B + {}^{1}_{1}H + 7\, {}^{1}_{0}n$$

A then emits a positron to become B, and B emits a positron to become ${}^{52}_{24}Cr$. Identify A and B and write equations for the various reactions.

Solution

$${}^{52}_{24}Cr + {}^{4}_{2}He \rightarrow {}^{52}_{26}Fe\ (A) + 4\, {}^{1}_{0}n$$

$${}^{52}_{24}Cr + {}^{4}_{2}He \rightarrow {}^{52}_{25}Mn\ (B) + {}^{1}_{1}H + 3\, {}^{1}_{0}n$$

$${}^{52}_{26}Fe \rightarrow {}^{52}_{25}Mn + {}^{0}_{1}e$$

$${}^{52}_{25}Mn \rightarrow {}^{52}_{24}Cr + {}^{0}_{1}e$$

On a problem of this type, the easy way to proceed is to start at the end and work backwards. Try it.

All elements beyond atomic number 83 are naturally radioactive and decay by a variety of alpha and beta emissions. A good example is one of the so-called natural radioactive decay series.

Example 4.

${}^{235}_{92}U$ decays to ${}^{207}_{82}Pb$ by the following steps: $\alpha, \beta^-, \alpha, \beta^-, \alpha, \alpha, \alpha, \alpha, \beta^-, \alpha, \beta^-$. List the products formed in each successive step.

Solution.

Equations for the first three steps are

$${}^{235}_{92}U \rightarrow {}^{231}_{90}Th + {}^{4}_{2}He$$

$${}^{231}_{90}Th \rightarrow {}^{231}_{91}Pa + {}^{0}_{-1}e$$

$${}^{231}_{91}Pa \rightarrow {}^{227}_{89}Ac + {}^{4}_{2}He$$

Write equations for the remaining steps. The products are ${}^{227}_{90}Th$, ${}^{223}_{88}Ra$, ${}^{219}_{86}Rn$, ${}^{215}_{84}Po$, ${}^{211}_{82}Pb$, ${}^{211}_{83}Bi$, ${}^{207}_{81}T1$, and finally, ${}^{207}_{82}Pb$.

RATES OF NUCLEAR REACTIONS CATEGORY II

As chemistry students begin to learn about the periodic table, some obvious questions arise. Is Fr the least electronegative element? Is HAt a strong acid? Why aren't elements like Tc, Po, At, Rn, Fr, and Ra discussed along with other members of their families? The answer is that all of the known isotopes of these elements are radioactive with short half-lives. The *half-life* is the time required for ½ of any amount of a radioactive substance to decay into something else. The half-lives of the most stable isotopes of the elements just listed are ${}^{99}_{43}Tc$ (2.12 ×

10^5 yrs), $^{209}_{84}$Po ($\sim$100 yrs), $^{210}_{85}$At (8.3 hrs), $^{222}_{86}$Rn (3.825 days), $^{212}_{87}$Fr (19.3 min), and $^{226}_{88}$Ra (1622 years). None of these elements occur in nature. Even Tc with its relatively long half-life of 212,000 years would have disappeared from the earth's crust millions of years ago. In recent years, workable quantities of ^{99}Tc and ^{226}Ra have been artificially produced and much of the chemistry of these elements is known. Many chemical properties of the other elements are not well established, for obvious reasons. It would take a "fast" chemist to perform any kind of experiment with Fr.

As we can see from the few examples given, half-lives of radioactive substances vary considerably. Yet, the rate of decay of any radioactive element is a first order process (see Chapter 9) and obeys the equation

$$\log \frac{X_o}{X} = \frac{kt}{2.30}$$

Where X_o is the amount of material at some arbitrary initial time, X is the amount remaining after a time t, and K is the radioactive decay constant (1st order rate constant) for the specific substance decaying. We can solve this equation for the half-life by setting $X_o = 2$ and $X = 1$:

$$\log \frac{2}{1} = 0.301 = \frac{kt}{2.30}; k = \frac{0.692}{t_{1/2}}$$

Thus, we see that the half-life depends only on the specific radioactive decay constant of a substance and not upon the amount of substance. Any time we know k we can find $t_{1/2}$, and vice-versa.

Example 1.

Isotope $^{131}_{53}$I, used in medicine to analyze thyroid malfunction, has a radioactive decay constant of 8.60×10^{-2} (days)$^{-1}$. Find the half-life of $^{131}_{53}$I.

Solution.

$$t_{1/2} = \frac{0.692}{k} = \frac{0.692}{8.60 \times 10^{-2} \text{ (days)}^{-1}} = 8.05 \text{ days}$$

Example 2.

The isotope $^{40}_{19}$K occurs in nature since its half-life is quite long (1.2×10^9 yrs). It decays by beta emission to give $^{40}_{20}$Ca and by electron capture to give $^{40}_{18}$Ar. On earth, the ratio of ^{40}K to ^{40}Ar affords a reliable estimate of the amount of $^{40}_{19}$K that has decayed in any particular sample. Some rocks from the lunar surface contained ^{40}K and ^{40}Ar, and analyses indicated that 91% of the $^{40}_{19}$K had decayed. Estimate the age of the moon rocks.

Solution.

First, let's find the radioactive decay constant for $^{40}_{19}$K:

$$k = \frac{0.692}{t_{1/2}} = \frac{0.692}{1.2 \times 10^9 \text{ yrs}} = 5.8 \times 10^{-10} \text{ (yrs)}^{-1}$$

Then, we can use the decay rate equation. If 91% of the $^{40}_{19}K$ has decayed, 9% must remain, and the ratio X_0/X must be 1.00/.09. Thus,

$$\log \frac{X_0}{X} = \frac{kt}{2.30} \; ; \log \frac{1.00}{0.09} = \frac{5.8 \times 10^{-10} \; (yrs)^{-1} \times t}{2.30} = 1.046$$

and,

$$t = \frac{1.046 \times 2.30}{5.8 \times 10^{-10} \; (yrs)^{-1}} = 4.1 \times 10^9 \; yrs$$

Example 3.

The half-life of $^{210}_{88}At$ is 8.3 hours. If a chemist starts with a 10.0 mg sample of this isotope, how much is left after (a) 2.0 hrs? (b) 10.0 hrs? (c) two days?

Solution.

The radioactive rate constant is

$$k = \frac{0.692}{t_{1/2}} = \frac{0.692}{8.3 \; hrs} = 8.34 \times 10^{-2} \; (hrs)^{-1}$$

We can now substitute into the decay rate equation and solve for X for each of the different time periods.

(a) $$\log \frac{X_0}{X} = \frac{kt}{2.30} \; ; \log \frac{10.0 \; mg}{X} = \frac{(8.34 \times 10^{-2} \; (hrs)^{-1}) \times 2.0 \; hrs}{2.30}$$

$$\log \frac{10.0 \; mg}{X} = 0.0725 \; ; \frac{10.0 \; mg}{X} = 1.182$$

$$X = 8.46 \; mg \; after \; 2.0 \; hrs$$

Why can we use mg directly rather than converting to moles or numbers of atoms?

(b) $$\log \frac{10.0 \; mg}{X} = \frac{(8.34 \times 10^{-2} \; (hrs)^{-1}) \times 10.0 \; hrs}{2.30}$$

$$\log \frac{10.0 \; mg}{X} = 0.363 \; ; \frac{10.0 \; mg}{X} = 2.31$$

$$X = 4.33 \; mg \; after \; 10.0 \; hrs$$

(c) $$\log \frac{10.0 \; mg}{X} = \frac{8.34 \times 10^{-2} \; (hrs)^{-1} \times 48.0 \; hrs}{2.30}$$

$$\log \frac{10.0 \; mg}{X} = 1.741 \; ; \frac{10.0 \; mg}{X} = 5.51 \times 10^1$$

$$X = 0.181 \; mg \; after \; 48.0 \; hrs$$

Example 4.

The half-life of $^{212}_{87}Fr$ is 19.3 minutes. How long would it take for 3/4 of a sample of this isotope to decay?

Solution.

We could just reason this out. One half of the sample would be gone in 19.3 min. One half of the remaining sample would decay in another 19.3 min. Therefore, 3/4 of the original sample would have disappeared in 38.6 min.

Or, we could solve mathematically. The decay constant is $k = 0.692/19.3$ min $= 3.59 \times 10^{-2}$ $(min)^{-1}$, and if 3/4 of the sample has decayed, $X_o/X = 4/1$.

$$\log \frac{4}{1} = \frac{3.59 \times 10^{-2} \ (min)^{-1} \times t}{2.30} = 0.602$$

and

$$t = \frac{(0.602)(2.30)}{3.59 \times 10^{-2} \ (min)^{-1}} = 38.6 \text{ minutes}$$

Example 5.

A sample containing only $^{244}_{95}Am$ as a radioactive source gave a reading of 6500 counts per minute in a scintillation counter. After 1.0 hr, the reading had dropped to 1300 counts per minute. Find the decay constant and half-life of $^{244}_{95}Am$.

Solution.

Assuming the scintillation counter reading is directly proportional to the amount of emitting material, we can substitute into the decay rate equation and solve for k:

$$\log \frac{X_o}{X} = \frac{kt}{2.30} \ ; \ \log \frac{6500}{1300} = \frac{k \times 60 \ min}{2.30} = 0.699$$

$$k = \frac{(0.699)(2.30)}{60 \ min} = 0.0268 \ (min)^{-1}$$

Then,

$$t_{1/2} = \frac{0.692}{0.0268 \ (min)^{-1}} = 25.8 \text{ minutes}$$

CATEGORY III   MASS-ENERGY RELATIONS

Just as the nature and mechanisms of nuclear reactions are quite different from chemical reactions, so, too, are the energies of nuclear reactions quite different from the energies of chemical reactions. For example, the splitting of one mole of H_2 molecules into H atoms ($H_2 \rightarrow 2 H$) requires 103 kcal of energy, or conversely, 103 kcal is released when H atoms are combined to produce one mole of H_2. The splitting of one mole of He nuclei into protons and neutrons ($^4_2He \rightarrow 2 \, ^1_1H + 2 \, ^1_0n$) requires 629,000,000 kcal of energy, or, that much energy is released in the reverse process. Obviously, the forces involved must be quite different! The forces inherent in chemical bonds are believed to be electrostatic in nature. The forces inherent in nuclear binding are more difficult to visualize. We can, however, relate the energy change to the mass change in a nuclear process.

The relationship is concisely stated in the famous Einstein equation: $\Delta E = \Delta mc^2$, where ΔE is change in energy, Δm is change in mass, and c is a proportionality constant, the speed of light. When ΔE is kcal and Δm is in grams, the equation becomes

$$\Delta E \text{ (kcal)} = 2.15 \times 10^{10} \times \Delta m \text{ (grams)}$$

The energy which binds the nucleus together is appropriately called the *nuclear binding energy*. Nuclear binding energies may be calculated from the mass decrement, that is, the difference between the experimentally determined mass of the nucleus and the calculated mass based on the total number of protons and neutrons in the nucleus.

Example 1.

The observed masses of the deuterium nucleus, ^2_1H, and the lithium nucleus, ^7_3Li, are 2.01355 and 7.01436, respectively. Calculate the binding energy in kcal for
  (a)   one mole of ^2_1H nuclei
  (b)   one mole of ^7_3Li nuclei

Solution.

  (a)   The calculated mass of 1 mole of ^2_1H nuclei is

1 mole ^1_1H	=	1.00728 g
1 mole ^1_0n	=	1.00867 g
		2.01595 g
The observed mass	=	2.01355 g
The mass decrement, Δm	=	0.00240 g

Δm represents the mass that is "converted" to energy in binding together one mole of protons with one mole of neutrons. We can easily convert this to kcal of energy:

$$X \text{ kcal} = 0.00240 \text{ g} \times \frac{2.15 \times 10^{10} \text{ kcal}}{1 \text{ g}} = 5.16 \times 10^7 \text{ kcal}$$

Thus, to split apart one mole of ^2_1H into ^1_1H and ^1_0n would require 51,600,000 kcal.
  (b)   The calculated mass of a mole of ^7_3Li nuclei is

3 moles of protons	=	3×1.00728	=	3.02184 g
plus, 4 moles of neutrons	=	4×1.00867	=	4.03468 g
Total mass		=	7.05652 g
The observed mass		=	7.01436 g
Δm		=	0.04216 g

The binding energy is

$$X \text{ kcal} = 0.04216 \text{ g} \times \frac{2.15 \times 10^{10} \text{ kcal}}{1 \text{ g}} = 9.06 \times 10^8 \text{ kcal}$$

Average nuclear binding energies of the elements increase sharply from ^2_1H to $^{56}_{26}Fe$, then gradually decrease with increasing mass number. This phenomenon is the basis for the large amounts of energy released in nuclear fission and fusion reactions.

Example 2.

Calculate the energy change in kcal for

(a)   The fission of one mole of $^{239}_{94}Pu$ by the reaction

$$^{239}_{94}Pu + ^{1}_{0}n \rightarrow ^{90}_{38}Sr + ^{146}_{58}Ce + 4\,^{1}_{0}n + 2\,^{0}_{-1}e$$

(b)   The fusion of one mole of deuterium with one mole of tritium, $^{3}_{1}H$, by the reaction

$$^{2}_{1}H + ^{3}_{1}H \rightarrow ^{4}_{2}He + ^{1}_{0}n$$

The observed masses of the various isotopes are $^{239}_{94}Pu$ (239.0006), $^{90}_{38}Sr$ (89.8864), $^{146}_{58}Ce$ (145.8865), $^{2}_{1}H$ (2.01355), $^{3}_{1}H$ (3.01550), and $^{4}_{2}He$ (4.00150).

Solution.

Proceed by calculating the change in mass in going from reactants to products, then convert Δm to energy.

(a)

Product mass:	89.8864	$^{90}_{38}Sr$
+	145.8865	$^{146}_{58}Ce$
+ 4 × (1.00867)	4.03468	$4\,^{1}_{0}n$
+ 2 × (0.000549)	0.00110	$2\,^{0}_{-1}e$
	239.80868	
Reactant mass:	239.0006	$^{239}_{94}Pu$
+	1.00867	$1\,^{1}_{0}n$
	240.00927	

$$\Delta m = 240.00927 - 239.80868 = 0.20059 \text{ gram}$$

The mass units are grams since we are working on a mole scale. Note that there is a "loss" in mass in going from reactants to products. That mass is converted to energy that is released, i.e., the reaction is exothermic. The energy released is

$$X \text{ kcal} = 0.20059 \text{ g} \times \frac{2.15 \times 10^{10} \text{ kcal}}{1 \text{ g}} = 4.31 \times 10^9 \text{ kcal}$$

This, and/or similar reactions, take place in the plutonium atomic bomb.

(b)

Product mass:	4.00150	$^{4}_{2}He$
+	1.00867	$^{1}_{0}n$
	5.01017	
Reactant mass:	2.01355	$^{2}_{1}H$
+	3.01550	$^{3}_{1}H$
	5.02905	

$$\Delta m = 5.02905 - 5.01017 = 0.01888 \text{ gram per mole of } ^{4}_{2}He \text{ produced}$$

Converting to energy:

$$X \text{ kcal} = 0.01888 \text{ g} \times \frac{2.15 \times 10^{10} \text{ kcal}}{1 \text{ g}} = 4.06 \times 10^8 \text{ kcal}$$

This, and/or similar reactions, take place in what are popularly (or unpopularly!) called hydrogen bombs. While nuclear fission reactions can be controlled and the energy released slowly and steadily, the high temperatures (millions of $°C$) needed to initiate and sustain nuclear fusion have, so far, precluded the use of controlled nuclear fusion as a practical source of energy. In the hydrogen bomb, the high temperatures needed are produced by a fission bomb.

Example 3.

Calculate the energy changes, in kcal per mole, for the following nuclear reactions:

(a) $^{40}_{19}K \rightarrow ^{40}_{20}Ca + ^{0}_{-1}e$

(b) $^{63}_{30}Zn \rightarrow\rightarrow ^{63}_{29}Cu + ^{0}_{1}e$

(c) $^{210}_{84}Po \rightarrow ^{206}_{82}Pb + ^{4}_{2}He$

The isotopic masses needed are

$^{40}_{19}K$	39.95358	$^{63}_{30}Zn$	62.91665
$^{40}_{20}Ca$	39.95162	$^{63}_{29}Cu$	62.91305
$^{210}_{84}Po$	209.9368	$^{206}_{82}Pb$	205.9295
$^{4}_{2}He$	4.00150		

Solution.

Once again, the method is simply to calculate the change in mass for each reaction and convert to energy.

(a) Δm = (39.95162 + 0.000549) − 39.95358 = −0.00141 gram (a decrease in mass in the reaction)

$$X \text{ kcal} = 0.00141 \text{ g} \times \frac{2.15 \times 10^{10} \text{ kcal}}{1 \text{ g}} = 3.03 \times 10^{7} \text{ kcal}$$

(b) Δm = (62.91305 + 0.000549) − 62.91665 = −0.00305 gram

$$X \text{ kcal} = 0.00305 \text{ g} \times \frac{2.15 \times 10^{10} \text{ kcal}}{1 \text{ g}} = 6.56 \times 10^{7} \text{ kcal}$$

(c) Δm = (205.9295 + 4.00150) − 209.9368 = −0.0058 gram

$$X \text{ kcal} = 0.0058 \text{ g} \times \frac{2.15 \times 10^{10} \text{ kcal}}{1 \text{g}} = 1.25 \times 10^{8} \text{ kcal}$$

All these reactions are exothermic and spontaneous.

Category I   PROBLEMS

1. Write balanced nuclear equations for the following:
   (a) The decay of $^{36}_{17}Cl$ by beta emission.
   (b) The decay of $^{13}_{7}N$ by positron emission.
   (c) The decay of $^{230}_{90}Th$ by alpha emission.

2. Bombardment of $^{59}_{27}Co$ with deuterons produces ^{60}Co, which is one of several isotopes used in cancer treatment. ^{60}Co is a beta emitter with a half-life of 5.2 yrs. Write balanced equations for the production of ^{60}Co and for its beta decay.

3. Predict whether the following unstable nuclei would decay by beta or positron emission:
   (a) $^{9}_{3}Li$ (b) $^{20}_{11}Na$ (c) $^{35}_{18}Ar$ (d) $^{32}_{15}P$.

4. Write balanced nuclear equations for the following reactions.

Target	Bombarding Particle	Expelled Particle(s)
$^{37}_{17}Cl$	$^{2}_{1}H$	$^{4}_{2}He$
$^{208}_{82}Pb$	$^{4}_{2}He$	$2\,^{1}_{0}n$
$^{44}_{20}Ca$	$^{2}_{1}H$	$^{1}_{1}H$
$^{209}_{83}Bi$	$^{2}_{1}H$	$^{1}_{0}n$

5. Write plausible nuclear reactions for the conversion of $^{220}_{86}Rn$ to $^{208}_{82}Pb$.

Category II

6. A sample of $^{24}_{11}Na$ was placed in a scintillation counter that registered 15,000 counts per minute. After 5.0 hrs the count had dropped to 11,905 counts per minute. Find the half-life and decay constant for $^{24}_{11}Na$.

7. The radioactive isotope $^{14}_{6}C$ occurs naturally in the earth's biosphere as a result of cosmic radiation of $^{14}_{7}N$. The half-life of $^{14}_{6}C$ is 5760 years. All living organic matter contains a fixed amount of $^{14}_{6}C$, which diminishes when life ceases. A human bone from an Egyptian tomb contained only 59.55% of the $^{14}_{6}C$, found in living humans. Estimate the age of the bone.

8. The isotope $^{198}_{79}Au$ is used as a tracer in diagnosing liver disease. It decays by beta emission with a half-life of 2.69 days. How long would it take for 95% of a sample of $^{198}_{79}Au$ to decay?

9. Suppose rocks on the surface of Mars are found to have an $^{40}Ar/^{40}K$ ratio one half that of rocks found on earth. What could be implied by that? (The age of the earth is currently believed to be 4.5 billion years.) What is the most probable source of error in your interpretation?

10. One curie of a radioactive substance is the amount in which 3.7×10^{10} nuclei undergo decay in one second. Calculate the weight, in mg, of one curie of $^{210}_{84}Po$. ($t_{1/2} = 138$ days)

11. The half-life of tritium, $^{3}_{1}H$, is 12.46 years. Starting with a 2.00 g sample of tritium, how much would remain after
(a) 5 yrs?   (b) 15 yrs?   (c) 25 yrs?

Category III

12. Calculate the nuclear binding energies, in kcal/mole, for
(a) $^{4}_{2}He$ (b) $^{40}_{20}Ca$ (c) $^{206}_{82}Pb$.
(Refer to Example 3 for isotopic masses.)

13. Find the energy changes in kcal per mole for the following nuclear reactions:
(a) $^{3}_{1}H \rightarrow ^{3}_{2}He + ^{0}_{-1}e$
(b) $^{11}_{5}B + ^{2}_{1}H \rightarrow ^{11}_{6}C + 2\,^{1}_{0}n$
(c) $^{238}_{92}U \rightarrow ^{234}_{90}Th + ^{4}_{2}He$
Isotopic masses: $^{3}_{1}H$ 3.01550, $^{3}_{2}He$ 3.01493, $^{11}_{5}B$ 11.00656, $^{11}_{6}C$ 11.00814, $^{238}_{92}U$ 238.0003, $^{234}_{90}Th$ 233.9934.

14. The sun radiates energy at the rate of 9.56×10^{22} kcal per second.
(a) Assuming all solar energy is due to the fusion reaction $4\,^{1}_{1}H \rightarrow ^{4}_{2}He + 2\,^{0}_{1}e$, calculate the mass lost by the sun in one day.
(b) If the present mass of the sun is 2.0×10^{33} kg, how long will it take for one millionth of that mass to be converted to energy?

15. "Anti-matter" would consist of negative protons and positive electrons. If the reaction of anti-matter with matter resulted in 100% conversion of mass energy, calculate the energy that would be released if one mole of anti-hydrogen atoms reacted with one mole of normal hydrogen atoms.

General Problems

16. Complete the following:
    (a) $^{55}_{25}Mn +$ _____ $\rightarrow ^{55}_{26}Fe + 2\,^{1}_{0}n$
    (b) $^{6}_{3}Li + ^{1}_{0}n \rightarrow ^{3}_{1}H +$ _____
    (c) $^{14}_{7}N + ^{4}_{2}He \rightarrow ^{17}_{8}O +$ _____

17. In α decay the $^{4}_{2}He$ nuclei emitted pick up electrons from the surroundings and form helium gas. The isotope $^{212}_{83}Bi$, is an α emitter with a half-life of 60.0 min. Starting with 30.0 g of $^{212}_{83}Bi$, how much time would be required to produce 2.24 liters of He gas at 0°C and 1 atm pressure?

18. One of the fission products of the atomic bomb is $^{90}_{38}Sr$, which has a half-life of 29 years. If a soil sample contained 3.0×10^{-11} g of $^{90}_{38}Sr$ in 1955, how much will remain in the year 2000?

19. Calculate the energy change (in kcal per mole) for the reaction

$$^{7}_{3}Li + ^{2}_{1}H \rightarrow ^{9}_{4}Be$$

The appropriate masses are $^{7}_{3}Li$ (7.01436, $^{2}_{1}H$ (2.01355), $^{9}_{4}Be$ (9.00999).

20. From the information contained in (and your solution to) Problem 17, find the average number of α emissions per second that the $^{212}_{83}Bi$ undergoes to produce the 2.24 liters of helium gas.

PROBLEM SOLUTIONS

Chapter 1   THE FACTOR-LABEL METHOD

1. $X \text{ yards} = 100 \text{ meters} \times \dfrac{100 \text{ cm}}{1 \text{ meter}} \times \dfrac{1 \text{ in}}{2.54 \text{ cm}} \times \dfrac{1 \text{ yard}}{36.0 \text{ in}} = 109.4 \text{ yards}$

2. $X \text{ kg} = 5 \text{ lbs} \times \dfrac{454 \text{ g}}{1 \text{ lb}} \times \dfrac{1 \text{ kg}}{1000 \text{ g}} = 2.27 \text{ kg}$

3. $X \text{ lbs Hg} = 1 \text{ qt Hg} \times \dfrac{1 \text{ liter Hg}}{1.06 \text{ qt Hg}} \times \dfrac{1000 \text{ ml Hg}}{1 \text{ liter Hg}} \times \dfrac{13.6 \text{ g Hg}}{1 \text{ ml Hg}} \times \dfrac{1 \text{ lb Hg}}{454 \text{ g Hg}} = 28.3 \text{ lbs}$

4. $X \text{ min} = 9.3 \times 10^7 \text{ miles} \times \dfrac{5280 \text{ ft}}{1 \text{ mile}} \times \dfrac{12 \text{ in}}{1 \text{ ft}} \times \dfrac{2.54 \text{ cm}}{1 \text{ in}} \times \dfrac{1 \text{ sec}}{3.0 \times 10^{10} \text{ cm}} \times \dfrac{1 \text{ min}}{60 \text{ sec}} =$
8.3 min

5. $X \text{ hrs} = 200 \text{ km} \times \dfrac{1000 \text{ meters}}{1 \text{ km}} \times \dfrac{100 \text{ cm}}{1 \text{ meter}} \times \dfrac{1 \text{ in}}{2.54 \text{ cm}} \times \dfrac{1 \text{ ft}}{12 \text{ in}} \times \dfrac{1 \text{ mile}}{5280 \text{ ft}} \times \dfrac{1 \text{ hr}}{55 \text{ miles}}$
2.26 hrs

            (Knowing 1 km = 0.62 mi would make it easier.)

6. $X \$ = 1 \text{ lb Cf} \times \dfrac{454 \text{ g}}{1 \text{ lb}} \times \dfrac{\$100}{10^{-7} \text{ g}} = \$454,000,000,000$

7. $X \text{ in} = 20 \text{ mm} \times \dfrac{1 \text{ cm}}{10 \text{ mm}} \times \dfrac{1 \text{ in}}{2.54 \text{ cm}} = 0.787 \text{ in}$

8. $X \text{ CO}_2 \text{ molecules} = 1 \text{ mile} \times \dfrac{5280 \text{ ft}}{1 \text{ mile}} \times \dfrac{12 \text{ in}}{1 \text{ ft}} \times \dfrac{2.54 \text{ cm}}{1 \text{ in}} \times \dfrac{1 \text{ angstrom}}{10^{-8} \text{ cm}} \times \dfrac{1 \text{ CO}_2 \text{ molecule}}{5.0 \text{ angstroms}} =$
$3.2 \times 10^{12} \text{ CO}_2$ molecules

            (How many grams would that be? See Chapter 2.)

9. $X \dfrac{\text{meters}}{\text{sec}} = 55 \dfrac{\text{miles}}{\text{hrs}} \times \dfrac{1 \text{ km}}{0.62 \text{ mi}} \times \dfrac{1000 \text{ meters}}{1 \text{ km}} \times \dfrac{1 \text{ hr}}{60 \text{ min}} \times \dfrac{1 \text{ min}}{60 \text{ sec}} = 24.6 \text{ m/sec}$

10. $X \text{ lbs} = 10^{11} \text{ Pb atoms} \times \dfrac{207 \text{ g}}{6.02 \times 10^{23} \text{ Pb atoms}} \times \dfrac{1 \text{ lb}}{454 \text{ g}} = 7.57 \times 10^{-14} \text{ lbs}$

Chapter 2   MOLES, FORMULAS AND STOICHIOMETRY

Category I

1.   (a) $X \text{ moles Sn} = 17.5 \text{ g Sn} \times \dfrac{1 \text{ mole Sn}}{118.7 \text{ g Sn}} = 0.147$

     (b)   0.20   (c)   0.855   (d)   1.05

2. (a) $X \text{ g } CH_3OH = 0.20 \text{ mole } CH_3OH \times \dfrac{32.0 \text{ g } CH_3OH}{1 \text{ mole } CH_3OH} = 6.40$

    (b)   4.90   (c)   126   (d)   25.5

3. (a) $X \text{ molecules } C_6H_6 = 6.0 \text{ moles } C_6H_6 \times \dfrac{6.02 \times 10^{23} \text{ molecules}}{1 \text{ mole } C_6H_6} = 3.61 \times 10^{24}$

    (b) 4.62×10^{22} (c) 1.74×10^{22} (d)   there are no *molecules* of NaCl

4. (a) $X \text{ g Pb} = 1 \text{ atom Pb} \times \dfrac{207 \text{ g Pb}}{6.02 \times 10^{23} \text{ atoms Pb}} = 3.44 \times 10^{-22}$

    (b) 5.68×10^{-20} g   (c)   79,800 g   (d)   351 g

Category II

5. $X \text{ moles N atoms} = 25.0 \text{ g } N_2O_5 \times \dfrac{1 \text{ mole } N_2O_5}{108 \text{ g } N_2O_5} \times \dfrac{2 \text{ moles N atoms}}{1 \text{ mole } N_2O_5} = 0.463$

6. $X \text{ g O} = 4.0 \text{ g } C_3H_6O \times \dfrac{16.0 \text{ g O}}{58.0 \text{ g } C_3H_6O} = 1.10$

7. $X \text{ moles S} = 6.84 \text{ g } Al_2(SO_4)_3 \times \dfrac{1 \text{ mole } Al_2(SO_4)_3}{342 \text{ g } Al_2(SO_4)_3} \times \dfrac{3 \text{ moles S}}{1 \text{ mole } Al_2(SO_4)_3} = 0.06$

8. $X \text{ g M} = 1 \text{ mole M} \times \dfrac{2 \text{ moles O } (MO_2)}{1 \text{ mole M } (MO_2)} \times \dfrac{32.0 \text{ g O}}{2 \text{ moles O}} \times \dfrac{3.0 \text{ g M}}{6.0 \text{ g O}} = 16.0$

9. $X \text{ moles N} = 30.4 \text{ g N} \times \dfrac{1 \text{ mole N}}{14.0 \text{ g N}} = 2.17$

    $X \text{ moles O} = 69.6 \text{ g O} \times \dfrac{1 \text{ mole O}}{16.0 \text{ g O}} = 4.35$

    Ratio: O/N = 4.35/2.17 = 2/1. Formula = NO_2

10. moles Ca = 7.80 ÷ 40.1 = 0.195
     moles Cr = 20.3 ÷ 52.0 = 0.390
     moles O = 21.84 ÷ 16.0 = 1.365
     Ratio: Cr/Ca = 0.390/0.195 = 2/1
                  O/Ca = 1.365/0.195 = 7/1
     Formula = $CaCr_2O_7$

11. moles C = 92.3 ÷ 12.0 = 7.69
     moles H = 7.7 ÷ 1.0 = 7.7
     simplest formula = CH; CH (Wt 13) × 6 = 78
                  Molecular formula = C_6H_6

12. $\dfrac{207 \text{ g Pb}}{323 \text{ g Pb}(C_2H_5)_4} = .6409 \times 100 = 64.09\% \text{ Pb}$

Category III

13. (a) $3 NO_2 + H_2O \rightarrow 2 HNO_3 + NO$

    (b) $C_6H_6 + \dfrac{15}{2} O_2 \rightarrow 6 CO_2 + 3 H_2O$ or $2 C_6H_6 + 15 O_2 \rightarrow 12 CO_2 + 6 H_2O$

    (c) $Al_2O_3 + 3 H_2O \rightarrow 2 Al(OH)_3$

    (d) $C_3H_8O + \dfrac{9}{2} O_2 \rightarrow 3 CO_2 + 4 H_2O$ or $\times 2$

    (e) $2 NH_3 + \dfrac{5}{2} O_2 \rightarrow 2 NO + 3 H_2O$ or $\times 2$

14. X moles NO = 13.8 g NO_2 $\times \dfrac{1 \text{ mole } NO_2}{46.0 \text{ g } NO_2} \times \dfrac{1 \text{ mole } NO}{3 \text{ moles } NO_2}$ = 0.10

15. X g C_2H_5Br = 5.42 g PBr_3 $\times \dfrac{1 \text{ mole } PBr_3}{271.0 \text{ g } PBr_3} \times \dfrac{3 \text{ moles } C_2H_5Br}{1 \text{ mole } PBr_3} \times \dfrac{109.0 \text{ g } C_2H_5Br}{1 \text{ mole } C_2H_5Br}$ = 6.54 g

16. X moles C = 8.80 g CO_2 $\times \dfrac{1 \text{ mole } C}{44.0 \text{ g } CO_2}$ = 0.20

  X moles H = 5.40 g H_2O $\times \dfrac{2 \text{ moles } H}{18.0 \text{ g } H_2O}$ = 0.60

  H/C = 3/1; simplest formula = CH_3

17. $2\ CH_3NO_2 + \dfrac{5}{2}\ O_2 \rightarrow 2\ CO_2 + 3\ H_2O + 2\ NO$

  X g O_2 = 122 g CH_3NO_2 $\times \dfrac{1 \text{ mole } CH_3NO_2}{61.0 \text{ g } CH_3NO_2} \times \dfrac{5/2 \text{ moles } O_2}{2 \text{ moles } CH_3NO_2} \times \dfrac{32.0 \text{ g } O_2}{1 \text{ mole } O_2}$ = 80.0 g

18. $N_2 + 3\ H_2 \rightarrow 2\ NH_3$

  $\dfrac{50.0}{28} = 1.79$ moles N_2 ; $\dfrac{30.0}{2.0} = 15.0$ moles H_2

  H_2 is in large excess so reaction will be limited by amount of N_2.

  X g NH_3 = 1.79 moles N_2 $\times \dfrac{2 \text{ moles } NH_3}{1 \text{ mole } N_2} \times \dfrac{17.0 \text{ g } NH_3}{1 \text{ mole } NH_3}$ = 60.9 g

General Problems

19. (a) 10   (b) 2.3   (c) 2.5×10^{-13} (d) .017
  (e) 0.25

20. Mn_2O_3

21. 6.96 g

22. 52,800 g

23. 331 g

24. A = XeF_4 ; B = XeF_6

Chapter 3   THERMODYNAMICS

Category I

1. (a) cost = 225 kcal (N_2) + 3 $\times$ 103 kcal (H_2) = 534 kcal
    gain = 6 $\times$ 92 (N–H) = 552 kcal
    ΔH (reaction = –552 + 534 = –18 kcal
  (b) cost = 2 $\times$ 109 (H_2O) + 57 (Cl_2) = 275 kcal
    gain = 2 $\times$ 102 (HCl) + ½ $\times$ 116 (O_2) = 262 kcal
    ΔH (reaction) = –262 + 275 = +13 kcal
  (c) cost = 2 $\times$ 71 (HI) + 45 (Br_2) = 187 kcal
    gain = 2 $\times$ 87 (HBr) + 36 (I_2) = 210 kcal
    ΔH (reaction) = –210 + 187 = –23 kcal

2. cost = 3 $\times$ 92 (NH_3) + 3 $\times$ 36 (I_2) = 384 kcal
  gain = 3 $\times$ 71 (HI) + 3 $\times$ B (B = bond energy of N–I bond)
  ΔH (reaction) = –8.5 kcal = –(213 + 3B) + 384

  3B = 179.5 ; B = 59.8 kcal

156 **PROBLEM SOLUTIONS**

3. (a) cost = 36 (I_2) + 2 × 102 (HCl) = 240
gain = 2 × 71 (HI) + 57 (Cl_2) = 199
ΔH (reaction) = −199 + 240 = +41 kcal

(b) cost = (2 × 85)(diamond)* + 2 × 57 (Cl_2) = 284
gain = 4 × 78 (CCl_4) = 312
ΔH (reaction) = −312 + 284 = −28 kcal
*(each C atom *shares* 4 bonds)

(c) cost = 2 × 109 (H_2O) + 45 (Br_2) = 263
gain = 2 × 87 (HBr) + 58 $(\frac{1}{2}O_2)$ = 232
ΔH (reaction) = −232 + 263 = +31 kcal

4. $CH_4 + 2 O_2 \rightarrow CO_2 + 2 H_2O$
cost = 4 × 98 (CH_4) + 2 × 116 (O_2) = 624
gain = 2 × B (B = bond energy of C=O bond) + 4 × 109 (H_2O)
ΔH (reaction) = −gain + cost = cost − gain = −211 kcal = 624 − (2B + 436)

$$2B = 399 \; ; B = 199.5 \text{ kcal}$$

Category II

5. (a) ΔH (reaction) = (−94.1 + 0(H_2)) − (−26.4 + −57.8) = −9.9 kcal
(b) ΔH (reaction) = (−342.4 + −68.4 + −94.1) − (−288.5 + −216.9) = +0.50 kcal
(c) ΔH (reaction) = (0(I_2) + 2x − 68.4) − (−44.8 + 2x + 6.2) = −104.4 kcal

6. −52.2 = ΔH_f $(C_2H_4Cl_2)$ − (+12.5 + 0) ΔH_f = −39.7 kcal

7. (a) ΔH (reaction) = (−216.9) − (−94.5 + −68.4) = −54.0 kcal
(b) ΔH (reaction) = (−342.4) − (−151.9 + −94.5) = −96.0 kcal
(c) ΔH (reaction) = (−342.4 + −68.4) − (−151.9 + −216.9) = −42.0 kcal

8. $H_2SO_4 \rightarrow SO_3 + H_4O$ ΔH = +54.0 kcal
$CaO + SO_3 \rightarrow CaSO_4$ ΔH = −96.0 kcal
$CaO + H_2SO_4 \rightarrow CaSO_4 + H_2O$ ΔH = −42.0 kcal
Chemical equations and heat energies are additive.

9. +89.2 = (−26.4 + ΔH_f(NO)) − (−94.1 + 0) ΔH_f (NO) = +21.5 kcal

Category III

10. $\Delta H = C × (t_1 − t_2)$
X kcal = 1.14 g C_8H_{18} × $\dfrac{−1300 \text{ kcal}}{114 \text{ g } C_8H_{18}}$ = ΔH = −13.0 kcal
−13,000 cal = 1000 cal/°C (25.0°C − t_2) t_2 = 38.0°C

11. ΔH = 500 cal/°C (26.22°C − 28.54°C)
ΔH = −1160 cal

12. ΔH = 1.54 kcal/°C (25.0°C − 26.75°C) = −2.70 kcal
(3 figures) for oxidation of 1.04 g Cr
X kcal = 2.0 mole Cr × $\dfrac{52.0 \text{ g Cr}}{1.0 \text{ mole Cr}}$ × $\dfrac{2.70 \text{ kcal}}{1.04 \text{ g Cr}}$ = −270 kcal per mole of Cr_2O_3 formed

Category IV

13. $\Delta G = \Delta H − T \Delta S$ = −21.6 − (−.0023)(298) = −20.9 kcal

14. O = +21.6 − T (+.0023) ; T = 9391°K

15. (a) −
(b) +

(c) –
(d) +
(e) +
(f) near zero
(g) +

16. (a) $\Delta G = (-272.2) - (-126.3 + -94.3) = -51.6$ kcal
    (b) $\Delta G = (-88.5 + -56.7) - (-7.9 + 0) = -137.3$ kcal
    (c) $\Delta G = (+12.4) - (+20.7 + 0) = -8.3$ kcal
    (d) $\Delta G = (-45.1 + -94.3) - (-52.3 + -32.8) = -54.3$ kcal
    (e) $\Delta G = (0 + 2x - 12.7) - (-7.9 + 0) = -17.5$ kcal

17. $\Delta G = (-94.3) - (-32.8 + 0) = -61.5$ kcal
    $\Delta H = (-94.1) - (-26.4 + 0) = -67.7$ kcal
    $\Delta G = \Delta H - T\Delta S;\ -61.5 = -67.7 - (298°K)\Delta S;\ \Delta S = 0.0208$ kcal/°K

18. $\Delta S = \Delta H/T = 2180/703 = 3.10$ cal/mole-°K

19. ΔS (per mole) $= 9400/351 = 26.8$ cal/°K
    mw $C_2H_5OH = 46$, and 2.3 g$/46 = 0.05$ mole
    Hence, $0.05 \times 26.8 = 1.34$ cal/°K for 2.30 grams

20. (a) $\Delta G = \Delta H - T\Delta S = -67.7 - (2000°K)(-0.0208) = -26.1$ kcal
    (b) $\Delta G = -67.7 - (4000°K)(-0.0208) = +15.5$ kcal

General Problems

21. (a) Bond energies: +13 kcal
         Heats of formation: +13.6 kcal
    (b) Bond energies: –54 kcal
         Heats of formation: –70.2 kcal
    Bond energies based on gas phase. Reaction gives I_2(s).

22. –92.1 kcal/mole

23. (a) +4.25 kcal
    (b) –1.60 kcal

24. $\Delta G = -82.7$ kcal/mole
    $\Delta H = -96.0$ kcal/mole
    $\Delta S = -0.0446$ kcal/mole-°K

Chapter 4   GASES

Category I

1. One mole (44.0 g) CO_2 occupies 22.4 ℓ at STP, so 2.20 g (.05 mole) occupies $.05 \times 22.4$ = 1.12 liters.

2. $T = PV/nR = \dfrac{1.00 \text{ atm} \times 8.96 \text{ ℓ}}{0.20 \text{ mole} \times .0821 \text{ atm./ℓ/mole-°K}} = 546°K$

    or
    $= 273°K$

3. $V = nRT/P = \dfrac{(22.4/28) \text{ mole} \times 62.4 \text{ mm-ℓ/mole-°K} \times 300°K}{700 \text{ mm}}$
    $V = 21.4$ liters

4. X atm $= 7.6$ m bar $\times \dfrac{1 \text{ bar}}{1000 \text{ m bar}} \times \dfrac{0.987 \text{ atm}}{1 \text{ bar}} = 7.5 \times 10^{-3}$ atm

$$n/V = P/RT = \frac{7.5 \times 10^{-3} \text{ atm}}{0.0821 \text{ atm-}\ell/\text{mole-}^\circ K \times 250^\circ K} = 3.7 \times 10^{-4} \text{ mole of gas per liter}$$

5. $wt/V = P \times m.w./RT = (50 \times 44.0)/(.0821 \times 700) = 38 \text{ grams per liter}$

6. $P = \left(\dfrac{wt}{V}\right) \times \dfrac{RT}{mw} = 1.50 \times \dfrac{.0821 \times 300}{4.0} = 9.24 \text{ atm}$

7. $mw = \dfrac{wt \times RT}{PV} = \dfrac{(0.722 \text{ g})(62.4 \text{ mm-}\ell/\text{mole-}^\circ K)(363^\circ K)}{(740 \text{ mm})(0.250 \ell)} = 88.4 \text{ g}$

8. $2 \text{ HgO} \rightarrow 2\text{Hg}^\circ + O_2$. $10.83 \text{ g} \div 216.6 = 0.05$ mole HgO, which would yield 0.025 mole of O_2 gas. Then

$$V = \frac{nRT}{P} = \frac{0.025 \times 62.4 \times 333}{680} = 0.764 \text{ liter}$$

Category II

9. $V_2 = V_1 \times \dfrac{P_1}{P_2} \times \dfrac{T_2}{T_1} = 40,000 \ \ell \times \dfrac{760}{700} \times \dfrac{270}{303} = 38,700 \text{ liters}$

10. $P_2 = P_1 \times \dfrac{V_1}{V_2} \times \dfrac{T_2}{T_1} = 1.20 \text{ atm} \times 1 \times \dfrac{473}{293} = 1.94 \text{ atm}$

11. $V_2 = 5.0 \ \ell \times \dfrac{710 \text{ mm}}{1672 \text{ mm}} \times \dfrac{243^\circ K}{298^\circ K} = 1.73 \text{ liters}$

12. $T_2 = T_1 \times \dfrac{P_2}{P_1} \times \dfrac{V_2}{V_1} = 323^\circ K \times 1 \times \dfrac{7000 \ \ell}{6000 \ \ell} = 377^\circ K$

(assuming pressure remains constant)

Category III

13. From total pressure we found 3.7×10^{-4} mole per liter as the total gas concentration of the Martian atmosphere (Prob. 4). From Avogadro's principle, the molar concentration of O_2 would be $1\% \times 3.7 \times 10^{-4} = 3.7 \times 10^{-6}M$, and the partial pressure due to O_2 is

$$P = n/V \times RT = 3.7 \times 10^{-6} \times .0821 \times 250 = 7.6 \times 10^{-5} \text{ atm}$$

14. $20\% \times 1.20 \text{ atm} = 0.24 \text{ atm}$

15. $R_{CO_2} = R_{C_2F_6} \times \dfrac{\sqrt{138}}{\sqrt{44}}$; $R_{CO_2} = 1.77 \times R_{C_2F_6}$; $1.77 \times 26 \text{ hrs} = 46 \text{ hrs}$

16. $\sqrt{mw} \times = \sqrt{4.0} \times \dfrac{R_{He}}{R_x} = 2.0 \times \dfrac{42.33 \text{ hrs}}{26.33 \text{ hrs}} = 3.2$; $mw = 10.3$

(Note: rate is inversely proportional to time traveled.)

General Problems

17. $9.6 \text{ g}/\ell$

18. $36.0 \ \ell$

19. $273^\circ K \ (0^\circ C)$

20. 36

21. 1920 mm Hg

22. 31.2 ℓ

23. 7.1 atm

24. 305°K

25. 12.0 ℓ

Chapter 5   ATOMIC STRUCTURE AND BONDING

Category I

1. (a) $1s^2 2s^2 2p^5$ F
    (b) (Ne core) $3s^2 3p^2$ Si
    (c) (Ar core) $4s^1$ K
    (d) (Ar core) $3d^5 4s^2$ Mn
    (e) (Xe core) $4f^{14} 5d^7 6s^2$ Ir
    (f) (Kr core) $4d^{10} 5s^2 5p^4$ Te

2. (a) (Ne core) $3s^2 3p^6$ Cl^-
    (b) (Ne core) $3s^2 3p^6$ Ca^{2+}
    (c) (Ar core) $3d^6$ Fe^{2+}
    (d) (Ar core) $3d^3$ Cr^{3+}
    (e) $1s^2 2s^2 2p^6$ N^{3-}
    (f) (Xe core) $4f^{14} 5d^{10}$ Hg^{2+}

3. (a) Rb
    (b) As
    (c) Co^{2+}
    (d) Ar, K^+, Ca^{2+}, Sc^{3+}, Ti^{4+}, Cl^-, S^{2-}, P^{3-}
    (e) Zn, Ge^{2+}, As^{3+}, Se^{4+}
    (f) Ta

4. (a) O   (b) Ge   (c) Ti   (d) Bi   (e) Y

5. (Rn core) $5f^{14} 6d^{10} 7s^2 7p^6$

Category II

6. (a) H:C̈l:
    (b) :C:::O:
    (c) :N:::O:⁺
    (d) :C:::N:⁻
    (e) :N:::N:

7. (a) MgO   (b) K_2O (c) Al_2O_3 (d) Ca_3N_2 (e) Cs_2Se

8. (a) 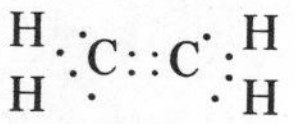 H ··   ·· H
        C::C
    H ··   ·· H

    (b) 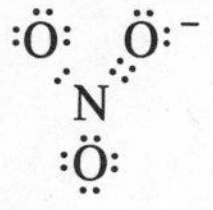 :Ö: Ö:⁻ $3 \times 6 = 18$ e⁻'s
           N   + 5
          :Ö: + 1
                  (extra) _____
                           24 e⁻'s total

    (c) 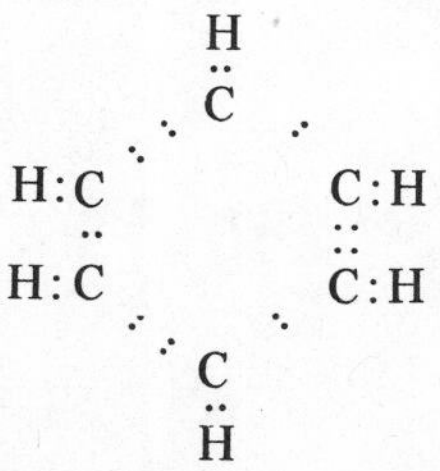   H
                     C
              · · · ·
           H:C   C:H
              ::   ::
           H:C   C:H
              · · · ·
                     C
                     H

    (d) H:S̈:C:::N: (H:S̈::C::N̈:)

    (e) 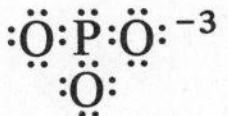 :Ö:P̈:Ö:⁻³   23 e⁻'s plus 3 "extra" = 26
              :Ö:

9.   (b), (c), and (d)

10.   (b), (c), and (e)

Category III

11.   (a)   Like H_2O in Example 3
      (b)   trigonal planar
      (c)   trigonal bipyramid

12.   See Example 3. Only one lone pair, so bond angle should be between 103° and 109°.

13.   Cl:Be:Cl for $BeCl_2$. Two lone pairs on $TeCl_2$, like H_2O in Example 3.

14.   Six e^- pairs, all bond pairs around S in SF_6. Seven e^- pairs around Xe, one lone pair that must "stick out" somewhere, distorting a perfect octahedron.

15.   (a)   like Example 2.

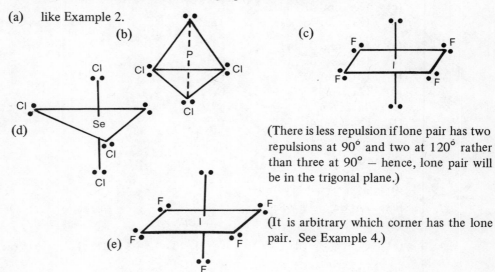

(There is less repulsion if lone pair has two repulsions at 90° and two at 120° rather than three at 90° — hence, lone pair will be in the trigonal plane.)

(It is arbitrary which corner has the lone pair. See Example 4.)

General Problems

16.   (a) XY_3 (b) XY_2

17.   (a)   V has an "Ar" structure in VF_5. VF_6 would necessitate breaking into a stable octet.
      (b)   Ti obeys the octet rule in $TiCl_4$ but not in $TiCl_3$.
      (c) C^{4-} and $(:C:::C:)^{2-}$ both obey the octet rule.

18.   (a)   (Kr core) $4d^{10}5s^25p^6$
      (b)   (Kr core) $4d^{10}4f^{14}5s^25p^65d^16s^2$
      (c)   (Kr core) $4d^{10}4f^{14}5s^25p^65d^4$
      (d)   (Rn core)

19.   (a) $\overset{..}{\underset{..}{F}} \quad \overset{..}{\underset{..}{F}} \quad \overset{..}{\underset{..}{F}}$
           $Al \quad Al$
           $\overset{..}{\underset{..}{F}} \quad \overset{..}{\underset{..}{F}} \quad \overset{..}{\underset{..}{F}}$
      (b) $:\overset{..}{\underset{..}{I}}:Pb:\overset{..}{\underset{..}{I}}:$
           $:\overset{..}{\underset{..}{I}}:$
      (c) $:\overset{..}{\underset{..}{Cl}}:$
           $:\overset{..}{\underset{..}{O}}:\overset{..}{S}:\overset{..}{\underset{..}{O}}:$
           $:\overset{..}{\underset{..}{O}}:$
      (d) $N::S:\overset{..}{\underset{..}{Cl}}:$

20.   (c) + (e)

21.   (a)   F   (b)   Kr   (c)   Rb   (d)   Ca

22.   Examples
      (a) H_2S (b) NF_3 (c) ICl_3 (d) CH_4

Chapter 6   LIQUIDS, SOLIDS AND PHASE CHANGES

Category I

1. $\log \dfrac{P_2}{P_1} = \dfrac{Hv}{4.58} \times \left(\dfrac{T_2 - T_1}{T_2 T_1} \right)$

    $\log P_2 - \log 760 = \dfrac{7170}{4.58} \times \left(\dfrac{298 - 350}{104300} \right)$

    $\log P_2 - 2.88 = -0.780;\ \log P_2 = 2.10;\ P_2 = 126$ mm hg

2. $\log \dfrac{10,040}{198} = \dfrac{\Delta Hv}{4.58} \left(\dfrac{300 - 200}{300 \times 200} \right)$; $\Delta Hv = 4686$ cal/mole

3. $\log \dfrac{450}{760} = \dfrac{9220}{4.58} \left(\dfrac{T_2 - 352}{T_2 \times 352} \right)$; $T_2 = 338°K = 65°C$

4. $\log \dfrac{5.7}{760} = \dfrac{9700}{4.58} \left(\dfrac{T_2 - 373}{T_2 \times 373} \right)$; $T_2 = 271°K = -2°C$

5. Find boiling point at 50 atm pressure.

    $\log \dfrac{50}{1} = \dfrac{9700}{4.58} \left(\dfrac{T_2 - 373}{T_2 \times 373} \right)$; $T_2 = 532°K$. Water would be vaporized at $700°K$.

Category II

6. (a) 1 at center + 1/8 $\times$ 8 at corners = 2
    (b) 1/8 $\times$ 8 at corners = 1

7. 3.52 Å

8. Volume = $(5.15 \times 10^{-8}$ cm$)^3 = 137 \times 10^{-24}$ cm^3

    mass of 4 Sr + 4 0 $= \dfrac{4(87.6 + 16.0)}{6.02 \times 10^{23}} = 68.8 \times 10^{-23}$ g

    density = m/v = 68.8×10^{-23} g $\div 137 \times 10^{-24}$ cm^3 = 5.02 g/cm^3

9. $(6.08)^2 + (6.08)^2 = \ell^2$, where ℓ = 4x radius of a Cs atom. ℓ = 8.60 Å; r = 2.15 Å

10. 1 center + ¼ $\times$ 4 "corners" = 2

Category III

11. ΔH_f (solid $\rightarrow$ liquid) = +1.44 kcal/mole
    ΔH_v (liquid $\rightarrow$ gas) = +9.72 kcal/mole
    $\Delta H_{sublimation}$ (solid $\rightarrow$ gas) = +1.44 + 9.72 = +11.16 kcal/mole

12. Substance sublimes when heated at constant 5 mm Hg pressure. Liquid solidifies when pressure increased from 50 mm Hg to 760 mm Hg at $60°C$.

13. (a) X cal = 15.6 g benzene $\times \dfrac{+32.7 \text{ cal}}{1 \text{ g benzene}}$ = + 510 cal

    (b) X cal = 15.6 g benzene $\times \dfrac{+94 \text{ cal}}{1 \text{ g benzene}}$ = +1466 cal

14. Melting 15.6 g of benzene will require 510 cal. Removing 510 cal from 100 g of water will lower the temperature by $5.1°C$. Final water temperature $\sim 15°C$.

15. For 1000 g H_2O ($25°C$) $\rightarrow$ 1000 g H_2O ($0°C$), 25,000 cal (25 kcal) would have to be removed. For 1000 g H_2O (liquid) $\rightarrow$ 1000 g H_2O (solid) at $0°C$: X cal = 1000 g H_2O $\times \dfrac{-80 \text{ cal}}{1 \text{ g } H_2O}$ = -80 kcal. Total heat removed = 25 kcal + 80 kcal = 105 kcal.

General Problems

16. $\Delta H_v = 7020$ cal/mole; $P_2 = 168$ mm Hg

17. 1080 atm

18. Assume corner atoms touching center atom. Length of unit cell = 4.30 Å; 2 atoms per unit cell; Avogadro's number = 5.96×10^{23}.

19. 4.23 Å

20. 248,000 cal

Chapter 7   CONCENTRATION

Category I

1. .25 mole/.250 ℓ = 1.0 M

2. 9.20 g/46 = 0.20 mole; 0.20 mole/1.50 ℓ = 0.133 M

3. 20.0 g CH_3OH = 20.0/32 = .625 mole CH_3OH
    80.0 g H_2O = 80.0/18 = 4.44 moles H_2O
    CH_3OH is the solute. Total volume of solution is

$$X \ \ell \text{ sol'n} = 100 \text{ g sol'n} \times \frac{.001 \ \ell \text{ sol'n}}{0.970 \text{ g sol'n}} = .103 \text{ liter}$$

$$X \text{ moles } CH_3OH = 1.0 \ \ell \text{ sol'n} \times \frac{.625 \text{ mole } CH_3OH}{.103 \ \ell \text{ sol'n}} = 6.07 \text{ M}$$

4. $X \text{ moles } HNO_3 = 1000 \text{ ml dil. sol'n} \times \dfrac{10.0 \text{ ml conc. sol'n}}{200 \text{ ml dil. sol'n}} \times \dfrac{18.0 \text{ moles } HNO_3}{1000 \text{ ml conc. sol'n}} = 0.90 \text{ M}$

Category II

5. $X \text{ moles } H_2SO_4 = 0.500 \ \ell \ H_2SO_4 \times \dfrac{1.5 \text{ moles } H_2SO_4}{1.0 \ \ell \ H_2SO_4} = 0.75 \text{ mole}$

6. $X \text{ moles } Fe(NO_3)_3 = 100 \text{ ml } Fe(NO_3)_3 \text{ sol'n} \times \dfrac{0.3 \text{ mole } Fe(NO_3)_3}{1000 \text{ ml } Fe(NO_3)_3} = 0.03 \text{ mole}$

7. $X \text{ g } C_6H_{12}O_6 = 150 \text{ ml sol'n} \times \dfrac{0.50 \text{ mole } C_6H_{12}O_6}{1000 \text{ ml sol'n}} \times \dfrac{180 \text{ g } C_6H_{12}O_6}{1.0 \text{ mole } C_6H_{12}O_6} = 13.5 \text{ g}$

8. $X \text{ g KOH} = 250 \text{ ml sol'n} \times \dfrac{0.10 \text{ mole KOH}}{1000 \text{ ml sol'n}} \times \dfrac{56.1 \text{ g KOH}}{1.0 \text{ mole KOH}} = 1.40 \text{ g}$

9. $X \text{ g A} = 40.0 \text{ ml dil sol'n} \times \dfrac{20.0 \text{ ml conc A}}{200 \text{ ml dil A}} \times \dfrac{1.0 \text{ mole A}}{1000 \text{ ml conc A}} \times \dfrac{89.0 \text{ g A}}{1 \text{ mole A } (C_3H_7O_2N)} =$
    0.356 g

Category III

10. (By definition) $\ell = \dfrac{.002 \text{ mole}}{0.15 \text{ M}} = 0.013 \ \ell$ or 13 ml

11. (By Factor-Label) $X \text{ mls sol'n} = 1.5 \times 10^{-4} \text{ mole Cl}^- \times \dfrac{1000 \text{ ml sol'n}}{0.40 \text{ mole Cl}^-} = 0.375 \text{ ml}$

12. $X \ \ell \text{ gas} = 1.0 \text{ g Pb} \times \dfrac{1 \text{ mole Pb}}{207 \text{ g Pb}} \times \dfrac{1.0 \ \ell \text{ gas}}{1.0 \times 10^{-6} \text{ mole Pb}} = 4.8 \times 10^3 \text{ liters gas}$

13. X moles $F^- = 1.0 \; \ell$ sol'n $\times \dfrac{1.0 \text{ g } F^-}{10^3 \; \ell \text{ sol'n}} \times \dfrac{1.0 \text{ mole } F^-}{19.0 \text{ g } F^-} = 5.3 \times 10^{-5}$ M

   $X \; \ell \; \text{``}H_2O\text{''} = 0.01 \text{ mole } F^- \times \dfrac{1.0 \; \ell \; \text{``}H_2O\text{''}}{5.3 \times 10^{-8} \text{ mole } F^-} = 1.9 \times 10^5$ liters

Category IV

14. X moles DDT = 1000 g benzene $\times \dfrac{3.55 \text{ g DDT}}{500 \text{ g benzene}} \times \dfrac{1 \text{ mole DDT}}{355 \text{ g DDT}} = 0.02$ m

15. moles $C_3H_6O = 0.50$ m $\times .250$ kg $H_2O = 0.125$. X g $C_3H_6O = 0.125$ mole $C_3H_6O \times$
   $\dfrac{58.0 \text{ g } C_3H_6O}{1.0 \text{ mole } C_3H_6O} = 7.25$ grams

16. X moles $H_2SO_3 = 1000$ ml sol'n $\times \dfrac{16.4 \text{ g } H_2SO_3}{500 \text{ ml sol'n}} \times \dfrac{1 \text{ mole } H_2SO_3}{82.1 \text{ g } H_2SO_3} = 0.40$
   Eq. wt. = 82.1/2 = 41.1 as an acid and as a reducing agent
   N = 2 $\times$ M = 0.80 N as acid and reducing agent

17. N $\times$ V (acid) = N $\times$ V (base)
   N (acid) $= \dfrac{0.20 \times 35.5}{50.0} = 0.142$ N

   X g acid = 1.0 equiv acid $\times \dfrac{1000 \text{ ml sol'n}}{0.142 \text{ equiv acid}} \times \dfrac{3.00 \text{ g acid}}{50.0 \text{ ml sol'n}} = 422.5$ g

18. Moles $H_2O = 24.3/18 = 1.35$; moles $CO_2 = 17.6/44 = 0.40$; moles $NO_2 = 8.70/46 = 0.19$;
   moles NO = 2.35/30 = 0.08

   mole fraction $NO_2 = \dfrac{0.19}{1.35 + 0.40 + 0.19 + 0.08} = 0.094$

Category V

19. mole fraction H_2O, $X = \dfrac{100.0/18.0}{100.0/18.0 + 80.0/46.0} = 0.762$
   $P_1 = X_1 P_1{}^0 = 0.762 \times 23.8$ mm = 18.1 mm

20. Δtb = B $\times$ m = 2.53 $\times$ 2.00 = 5.06 ; B.P. = 85.06°C

21. For 100.0 g lake water, 12.0 g $\div$ 58.5 = 0.205 mole NaCl and 88.0 g.
   H_2O molality $(Na^+ + Cl^-) = 0.410 \div 0.088$ kg H_2O = 4.66 m
   Δtf = F $\times$ m = 1.86 $\times$ 4.66 = 8.67°C ; F.P. = -8.67°C

22. m $= \dfrac{1.61}{1.86} = 0.866$ mole per 1000 g $H_2O \div 20 = 0.043$ mole per 50.0 g H_2O
   MW = wt/no. of moles = 2.60 g/.043 mole = 60.5

23. m (total particles) $= \dfrac{1.91}{1.86} = 1.03$ moles of HF, H^+, and F^- from 1.00 mole (per kg H_2O) of
   HF. If moles HF ionized = X, total moles = HF + H^+ + F^- = (1.00 - X) + X + X = 1.00 +
   X. So, X = moles HF ionized = 0.03 per 1.00 mole of HF dissolved per kg of H_2O.

General Problems

24.

Solute	Moles Solute	Grams Solute	Volume Solution	Molarity
HCl	0.15	5.48	0.30 ℓ	0.50
CH_3COOH	0.20	12.0	0.50	0.40
NH_3	0.80	13.6	4.0	0.20
$C_{12}H_{22}O_{11}$	0.02	6.84	100 ml	0.20

25.   2000 ml

26.   (a)   5.9 M   (b)   8.0 m   (c)   0.13

27.   (a)   1.82 M   (b)   2.00 m   (c)   1.22 g/ml

28.   (a)   0.062 M   (b)   0.124M   (c)   0.062 m   (d)   0.186 m   (e) $-0.346°C$

29.   0.356 M

30.   (a)   0.40 m   (b)   0.44 M   (c)   1.15 g/ml   (d)   raised by 2.01°C

Chapter 8   GENERAL EQUILIBRIUM (GASES)

Category I

1. $\dfrac{[PCl_3][Cl_2]}{[PCl_5]} = \dfrac{(0.20)(0.20)}{(0.80)} = 0.05$

2.   X moles CO $= 5.60$ g CO $\times \dfrac{1 \text{ mole CO}}{28.0 \text{ g CO}} = 0.20$

   X moles $O_2 = 3.20$ g $O_2 \times \dfrac{1 \text{ mole } O_2}{32.0 \text{ g } O_2} = 0.10$

   At equilibrium:

   X moles $CO_2 = 6.60$ g $CO_2 \times \dfrac{1 \text{ mole } CO_2}{44.0 \text{ g } CO_2} = 0.15$

   Thus,
   0.05 mole CO and 0.025 mole of O_2 left, and K is
   (remember, volume is 0.20 liter)

$$\dfrac{(.75)^2}{(.25)^2(.125)} = 72$$

3. $\dfrac{(X)(X)}{(2.0 - 2X)^2} = 0.016; \dfrac{X}{2.0 - 2X} = \sqrt{0.016}\,; X = 0.20$

4. $\dfrac{(X)(X)}{\sim(2.0)^2} = 2.5 \times 10^{-19}; X^2 = 10.0 \times 10^{-19}; X = 1.0 \times 10^{-9}$

5. $\dfrac{(5.6/5.0)(5.6/5.0)}{(0.20/5.0)} = K = 31.4$

6. $\dfrac{(X)(X)}{(1.0 - X)(2.0 - X)} = 3.0\,; 2.0\,X^2 - 9.0\,X + 6.0 = 0$

   $X = \dfrac{+9.0 \pm \sqrt{81.0 - 48.0}}{4.0} = 0.81 = [SO_3] = [NO]$

   $[SO_2] = 0.19\,; [NO_2] = 1.19$

7.   If $[H_2] = 6.0$, then $[N_2]$ would be 2.0, and finding $[NH_3]$ at equilibrium:

$$\dfrac{[NH_3]^2}{(2.0)(6.0)^3} = 1.8\,; [NH_3]^2 = 778\,; [NH_3] = 27.9$$

   Since 4.0 mole of NH_3 are "used up" reaching equilibrium, the answer is $27.9 + 4.0 = 31.9$.

Category II

8. (a) left   (b) right   (c) right   (d) right

9. $\dfrac{(10.0 + X)^2}{(3.0)^2 \, (1.02)} = 5.0 \times 10^3$; $X = 205$

10. $\dfrac{(0.10)\,(0.40)}{(0.60)(0.30)} = K = 0.22$; $\dfrac{(0.20)(0.50)}{(0.50)(0.30 + X)} = 0.22$; $X = 0.61$

11. $\dfrac{(2.0 + X)}{(1.0 - X)(1.0 - X)} = 4.0$; $4.0\,X^2 - 9.0\,X + 2.0 = 0$

    From quadratic formula, $X = 0.25$. So, at equilibrium, $[C_2H_2] = [H_2] = 0.75$, and $[C_2H_4] = 2.25$.

12. $\dfrac{(3.9)^2}{(2.0 - X)(2.5)} = 6.0$; $X = 0.99$

General Problems

13.

Change in Concentration	Equilibrium Concentration
−0.41	0.09
+0.41	0.41
+0.41	0.91

14. (a) right   (b) left   (c) right   (d) right

15. 0.092 mole

16. 2.50 liters

17. ΔS is positive and $T\Delta S$ becomes larger in the Gibbs-Helmholtz equation at higher temperatures.

Chapter 9   RATES OF REACTION

Category I

1. When the rate changes as a function of time, one can use the integrated form of the rate law(s) to determine reaction order. For a 1st order reaction, a plot of log X vs t will be linear with slope-k/2.3 and intercept at log X_0. For a second order reaction a plot of $1/X$ vs t will be linear with slope = k and intercept at $1/X_0$. Or, one can substitute X_0 and X into the equations and see which gives a constant value of k. Using either technique, we find this is a second order reaction:

$$\text{rate} = k(\text{conc. } NH_4CNO)^2; \quad k = 5.82 \times 10^{-2} \text{ M}^{-1}\text{-min}^{-1}$$

2. From 1st and 2nd experiments, (conc. H_3PO_4) and (conc. H^+) are held constant, (conc. I^-) is doubled, and the rate doubles. Therefore, reaction is first order in (conc. I^-). Similarly, from 1st and 3rd experiments, reaction is first order in H_3PO_4, and from 3rd and 4th, or 3rd and 5th experiments, reaction must be second order in H^+. Rate law is

$$\text{rate} = k(\text{conc. } H_3PO_4)(\text{conc. } I^-)(\text{conc. } H^+)^2$$

3. Since $t_{1/2}$ changes with X_0, assume 2nd order.

$$k = \frac{1}{t_{1/2}\,X_0} = \frac{1}{(200 \text{ sec})(0.01 \text{ M})} = \frac{1}{(66.7 \text{ sec})(0.03 \text{ M})} = 0.50 \text{ M}^{-1}\text{-sec}^{-1}$$

4. (a) For first order, $k = 0.693/t_{1/2} = 2.67 \times 10^{-2}$ min^{-1}

$$\log \frac{0.010 \text{ M}}{0.002 \text{ M}} = \frac{(2.67 \times 10^{-2} \text{ min}^{-1}) \times t}{2.30} \; ; \; t = 60.2 \text{ min}$$

(b) For second order, $k = \dfrac{1}{t_{1/2} \, X_0} = 3.85$ M^{-1} min^{-1}

$$\frac{1}{0.002 \text{ M}} - \frac{1}{0.010 \text{ M}} = (3.85 \text{ M}^{-1} \text{ min}^{-1}) \times t; \; t = 103.9 \text{ min}$$

5. $\log \dfrac{6.97 \times 10^2 \text{ sec}^{-1}}{4.30 \times 10^{-5} \text{ sec}^{-1}} = \dfrac{Ea}{4.58 \text{ cal/}^\circ\text{K}} \times \dfrac{500^\circ\text{K} - 300^\circ\text{K}}{500^\circ\text{K} \times 300^\circ\text{K}}$

$$Ea = 24{,}766 \text{ cal or } 24.8 \text{ kcal}$$

6. Second order in A, zero order in B

7. $k = 0.693/t_{1/2} = 9.72 \times 10^{-10}$ yrs^{-1}

$$\log \frac{1.00 \text{ g}}{X} = \frac{(9.72 \times 10^{-10} \text{ yrs}^{-1}) \times 4.5 \times 10^9 \text{ yrs}}{2.30} \; ; \; X = 0.013 \text{ g}$$

8. $\log \dfrac{9.60 \times 10^{-1}}{1.20 \times 10^{-1}} = \dfrac{12{,}000}{4.58} \times \dfrac{T_2 - 298}{T_2 \times 298}$

$$T_2 = 332^\circ\text{K or } 59^\circ\text{C}$$

Category II

9. Slowest step: $H_3PO_4 + I^- + 2 H^+ \rightarrow (H_3PO_4 \cdot I^- \cdot 2 H^+)^*$
    It would be very doubtful that a simultaneous 4 body collision is part of the mechanism.

10. $B_2 + C \rightarrow (B_2C)^*$ (slow)
    $(B_2C)^* \rightarrow X$
    $A + X \rightarrow AB + B + C$ (fast)

11. rate = k(conc. A)(conc. B)^2

12. $[Co(en)_2 Cl_2]^+ \rightarrow [Co(en)_2 Cl]^{2+} + Cl^-$ (slow)
    $[Co(en)_2 Cl]^{2+} + Br^- \rightarrow [Co(en)_2 BrCl]^+$ (fast)

General Problems

13. 1.62×10^{-3}, 3.24×10^{-3}, and 1.30×10^{-2} M/sec

14. $t_{1/2} = 2.0$ hrs; $k = 0.35$ hrs^{-1}

15. second order; $k = 3.20 \times 10^{-4}$ M^{-1} - min^{-1}

16. $k = 27$ sec^{-1}

17. $K = \dfrac{k_f}{k_r}$

Chapter 10   SOLUBILITY EQUILIBRIA

Category I

1. $X^2 = 3.0 \times 10^{-8}$; $X = [Zn^{2+}] = 1.73 \times 10^{-4}$

2. $X^2 = 3.0 \times 10^{-10}$; $X = [CrO_4{}^{2-}] = 1.73 \times 10^{-5}$

3. $Ag_2CO_3 \rightleftarrows 2 Ag^+ + CO_3{}^{2-}$ $\qquad \begin{cases} (X)^2(0.5 X) = 6.2 \times 10^{-12} \\ 0.5 X^3 = 6.2 \times 10^{-12} \end{cases}$
   $\qquad\qquad\qquad X \quad 0.5X$
   $[Ag^+] = X = 2.31 \times 10^{-4}$

4. $[Sr^{2+}] [CO_3{}^{2-}] = (4.0 \times 10^{-5})(4.0 \times 10^{-5}) = 1.60 \times 10^{-9}$

5. $Mg(OH)_{2(s)} \rightleftarrows Mg^{2+} + 2 OH^-$; $[OH^-] = 4.0 \times 10^{-4}$, $[Mg^{2+}] = 2.0 \times 10^{-4}$
   $[Mg^{2+}] [OH^-]^2 = (2.0 \times 10^{-4})(4.0 \times 10^{-4})^2 = 3.20 \times 10^{-11}$

Category II

6. $[Mg^{2+}](0.001)^2 = 3.2 \times 10^{-11}$; $[Mg^{2+}] = 3.2 \times 10^{-5}$

7. $(0.05)[SO_4{}^{2-}] = 6.4 \times 10^{-5}$; $[SO_4{}^{2-}] = 1.28 \times 10^{-4}$

8. $(0.01)[F^-]^2 = 1.7 \times 10^{-6}$; $[F^-] = 1.3 \times 10^{-2}$ without precipitation
   $X \text{ g NaF} = 100 \text{ ml sol'n} \times \dfrac{1.3 \times 10^{-2} \text{ mole NaF}}{1000 \text{ ml sol'n}} \times \dfrac{42.0 \text{ g NaF}}{1 \text{ mole NaF}} = 5.46 \times 10^{-2} \text{ g NaF for}$
   100 ml

9. $X \text{ moles Hg}^{2+} = 1.0 \text{ g HgCl}_2 \times \dfrac{1 \text{ mole Hg}^{2+}}{271.6 \text{ g HgCl}} = 3.7 \times 10^{-3}$

   $[Hg^{2+}] = 3.7 \times 10^{-3}$ mole/10,000 liters $= 3.7 \times 10^{-7}$ M. So, $[S^{2-}]$ could be $1.0 \times 10^{-50}/3.7 \times 10^{-7} = 2.7 \times 10^{-44}$ M without precipitation. In 10,000 liters, this is $10^4 \times 2.7 \times 10^{-44} = 2.7 \times 10^{-40}$ mole.

Category III

10. $0.001 \text{ M Ag}^+ \times 0.500 \text{ } \ell = 5.0 \times 10^{-4} \text{ mole Ag}^+$
    $0.001 \text{ M SCN}^- \times 0.500 \text{ } \ell = 5.0 \times 10^{-4} \text{ mole SCN}^-$
    Both are in total volume of 1.0 liter, so ion product is

    $$[Ag^+] [SCN^-] = (5.0 \times 10^{-4})(5.0 \times 10^{-4}) = 2.5 \times 10^{-7}$$

    This exceeds Ksp, so AgSCN will precipitate.

11. $[Ag^+] [BrO_3{}^-] = (5.0 \times 10^{-4})(5.0 \times 10^{-4}) = 2.5 \times 10^{-7}$. Ion product is less than Ksp (5.0×10^{-5}), so $AgBrO_3$ will not precipitate.

12. $0.01 \text{ M Hg}^{2+} \times 0.010 \text{ } \ell = 1.0 \times 10^{-4} \text{ mole Hg}^{2+} \div 1000 \text{ } \ell = 1.0 \times 10^{-7} \text{ M} = [Hg^{2+}]$. $[S^{2-}] = 0.001 \div 1000 \text{ } \ell = 1.0 \times 10^{-6}$. Ion product $= (1.0 \times 10^{-7}) \times (1.0 \times 10^{-6}) = 1.0 \times 10^{-13}$. Ksp $= 1.0 \times 10^{-50}$, so HgS would precipitate.

13. Add Ag^+ until $[Ag^+] = 1.0 \times 10^{-12}$. At that point $[I^-] = 1.0 \times 10^{-4}$ while conc. of Br^- and Cl^- remain 0.10 M.

14. Since $[M^{2+}] = 1.0$ M, ion products, $[M^{2+}] \times [S^{2-}]$ just equal $[S^{2-}]$. (a) and (b) all precipitate. (c) Mn^{2+} in solution, all others precipitate. (d) Mn^{2+} and Ni^{2+} in solution, Cu^+ and Pb^{2+} precipitate. (e) All but Cu^+ in solution. $[Cu^+]^2[S^{2-}] = (1.0)^2(1.0 \times 10^{-36}) = 1.0 \times 10^{-36}$, still exceeds Ksp.
    $[Ni^{2+}] [S^{2-}] = 1.4 \times 10^{-24}$; $(1.0)(S^{2-}) = 1.4 \times 10^{-24}$
    $[Pb^{2+}](1.4 \times 10^{-24}) = 4.2 \times 10^{-28}$; $[Pb^{2+}] = 3.0 \times 10^{-4}$
    At $[S^{2-}] = 1.4 \times 10^{-24}$, $[Ni^{2+}]$ still $= 1.0$, but $[Pb^{2+}]$ would drop to 0.0003 M.

General Problems

15. 2.7×10^{-23}

16. $[F^-] = 2.0 \times 10^{-3}$ M

17. $[SO_4{}^{2-}] = 1.1 \times 10^{-7}$ M

18. 6.6×10^{-5}

19. $[C_2H_3O_2{}^-] = 0.01$ M; $[SO_4{}^{2-}] = 0.01$ M

Chapter 11   ACID-BASE EQUILIBRIA

Category I

1.   (a) $[OH^- = \dfrac{kw}{[H^+]} = \dfrac{1.0 \times 10^{-14}}{2.6 \times 10^{-4}} = 3.8 \times 10^{-11}$

    (b) 1.15×10^{-6} (c) 1.0×10^{-14} (d) 2.3×10^{-9} (e) 1.50×10^{-3}

2.   If $[H^+] = A \times 10^{-B}$, pH = B − log A
    (a)   pH = 4 − log 2.6 = 4 − .41 = 3.59
    (b)   8.06   (c)   0   (d)   5.37   (e)   11.18

3.   (a) $[H^+] = Kw/.012 = 8.33 \times 10^{-13}$; pH = 12.08
    (b)   pH = 7.0   (c) $[H^+] = 7.5 \times 10^{-3}$; pH = 2.12
    (d)   0.50 g/56.1 = 8.9×10^{-3} mole KOH (or OH^-) ÷ 5.0 liters = 1.8×10^{-3}M = $[OH^-]$,
          $[H^+] = 5.6 \times 10^{-12}$ and pH = 11.25
    (e)   0.25 M × .010 ℓ = .0025 mole HC1 ÷ 0.300 ℓ = $[H^+] = 8.3 \times 10^{-3}$; pH = 2.08

4.   If pH = B − log A; $[H^+] = A \times 10^{-B}$
    (a) 2.0×10^{-5} (b) 4.8×10^{-10} (c)   10.0   (d) 5.0×10^{-8}
    (e) 6.5×10^{-6}

5.

	$[H^+]$	$[OH^-]$	pH
(a)	5.0×10^{-11}	2.0×10^{-4}	10.30
(b)	1.6×10^{-9}	6.3×10^{-6}	8.80
(c)	2.0×10^{-4}	5.0×10^{-11}	3.70
(d)	6.45×10^{-5}	1.55×10^{-10}	4.19
(e)	3.55×10^{-7}	2.82×10^{-8}	6.45

Category II

6. $\dfrac{(X)(X)}{1.0} = 3.2 \times 10^{-8}$; X = $[H^+] = 1.78 \times 10^{-4}$; pH = 3.75

7. $\dfrac{(X)(X)}{.50/.250} = 7.0 \times 10^{-4}$; X = $[H^+] = 3.74 \times 10^{-2}$

8. $\dfrac{[H^+](0.20)}{(0.50)} = 6.6 \times 10^{-5}$; $[H^+] = 1.65 \times 10^{-4}$; pH = 3.78

9.   pH = 2.42; $[H^+] = 3.80 \times 10^{-3}$; $\dfrac{(3.80 \times 10^{-3})(3.80 \times 10^{-3})}{(1.0)} = 1.44 \times 10^{-5}$

10. $\dfrac{(X)(X)}{(0.50 - X)} = 1.2 \times 10^{-2}$; $X = \dfrac{-.012 \pm \sqrt{1.44 \times 10^{-4} - (-.024)}}{2}$
                              X = $[H^+] = 7.2 \times 10^{-2}$

11. $\dfrac{(0.50)[Ac^-]}{(1.0)} = 1.8 \times 10^{-5}$; $[Ac^-] = 3.6 \times 10^{-5}$

Category III

12. (a) acid   (b) base   (c) base   (d) not hydrolyze
    (e) acid   (f) not hydrolyze   (g) hydrolyzes both ways, but NH_4OH
    is a stronger base than HCN is an acid, so, solution would be slightly basic.

13. $NH_4^+ + HOH \rightleftarrows NH_3 + H_3O^+$ $K_a = 5.6 \times 10^{-10}$

$$\frac{X^2}{2.0} = 5.6 \times 10^{-10}; X = [H^+] = 3.3 \times 10^{-5}; pH = 4.48$$

14. $Ag^+ + HOH \rightleftarrows AgOH_{(s)} + H^+$

$$\frac{[AgOH][H^+]}{[Ag^+]} = K_a; \text{ actually } \frac{[H^+]}{[Ag^+]} = K_a, \text{ since AgOH is solid. } \frac{(1.4 \times 10^{-5})}{(0.1)} = 1.4 \times 10^{-4}$$

Category IV

15. X moles NaOH = 100 ml $HNO_3 \times \dfrac{0.25 \text{ mole } HNO_3}{1000 \text{ ml } HNO_3} \times \dfrac{1 \text{ mole NaOH}}{1 \text{ mole } HNO_3} = 2.5 \times 10^{-2}$ mole

16. X moles NaOH = 100 ml $H_3PO_4 \times \dfrac{0.50 \text{ mole } H_3PO_4}{1000 \text{ ml } H_3PO_4} \times \dfrac{3 \text{ moles NaOH}}{1 \text{ mole } H_3PO_4} = 0.15$ mole

17.

mls HCl added	$[OH^-]$	$[H^+]$	pH
0	0.10	1.0×10^{-13}	13.0
20.0	4.29×10^{-2}	2.33×10^{-13}	12.63
40.0	1.11×10^{-2}	9.01×10^{-13}	12.05
49.0	1.01×10^{-3}	9.90×10^{-12}	11.00(4)
49.9	1.00×10^{-4}	1.00×10^{-10}	10.00
50.0	1.00×10^{-7}	1.00×10^{-7}	7.00
50.1	1.00×10^{-10}	1.00×10^{-4}	4.00
51.0	1.01×10^{-11}	9.90×10^{-4}	3.00(4)
60.0	1.10×10^{-12}	9.09×10^{-3}	2.04

18. By neutralizing the formic acid: $HCHO_2 + NaHCO_3 \rightarrow NaCHO_2 + \text{``} H_2CO_3 \text{''} (CO_2 + H_2O)$

19. X g $Ca(OH)_2$ = 10.0 g $HBr \times \dfrac{1 \text{ mole HBr}}{80.9 \text{ g HBr}} \times \dfrac{1 \text{ mole } Ca(OH)_2}{2 \text{ moles HBr}} \times \dfrac{74.1 \text{ g } Ca(OH)_2}{1 \text{ mole } Ca(OH)_2} = 4.58$ g

Category V

20. Only (c) and (d)

21. pH = 4.05; $[H^+] = 8.9 \times 10^{-5}$

$$\frac{[H^+][Ac^-]}{[HAc]} = 1.8 \times 10^{-5}; \frac{(8.9 \times 10^{-5})[Ac^-]}{(0.50)} = 1.8 \times 10^{-5}$$

$[Ac^-]$ = 0.10 M, or .025 mole Ac^- in .250 liter. If approximately all Ac^- comes from NaAc, then .025 mole of NaAc = (.025 × F.W. of $NaC_2H_3O_2$) = .025 × 82.0 = 2.05 g. By factor label:

$$X \text{ g NaAc} = .025 \text{ mole } Ac^- \times \frac{1 \text{ mole NaAc}}{1 \text{ mole } Ac^-} \times \frac{82.0 \text{ g NaAc}}{1 \text{ mole NaAc}} = 2.05 \text{ g}$$

22. (a) $\dfrac{[NH_4^+][OH^-]}{(NH_3)} = 1.8 \times 10^{-5}$: for $[NH_4^+] = [NH_3], [OH^-] = 1.8 \times 10^{-5}$;

$[H^+] = 1 \times 10^{-14}/1.8 \times 10^{-5} = 5.6 \times 10^{-10}$ and pH = 10 − log 5.6 = 9.25

(b)   0.05 mole of HCl neutralizes 0.05 mole of NH_3 and produces an additional 0.05 mole of NH_4^+, so: $[NH_3]$ = 0.15 mole/.100 liter = 1.50, and $[NH_4^+]$ = 0.25 mole/.100 liter = 2.50. Then,

$$\frac{(2.50)[OH^-]}{(1.50)} = 1.8 \times 10^{-5}; \quad [OH^-] = 1.08 \times 10^{-5}; \quad [H^+] = 9.3 \times 10^{-10}$$

and, pH = 9.03

(c)   Now, $[NH_4^+]$ goes to 1.50 M and $[NH_3]$ becomes 2.50 M. $[OH^-] = 3.01 \times 10^{-5}$; $[H^+] = 3.3 \times 10^{-10}$; pH = 9.48

23.   1.0 mole HF + 0.20 mole NaOH → 0.20 mole F^-

$$\frac{[H^+](1.20)}{(0.80)} = 7.0 \times 10^{-4}; \quad [H^+] = 4.67 \times 10^{-4}$$

General Problems

24. $CH_3NH_2 + HOH \rightleftarrows CH_3NH_3^+ + OH^-$; $K_b = 5.0 \times 10^{-4}$

25.   (a) 2.0×10^{-11} (b) 3.2×10^6 (c) 2.2×10^{-11}

26.   0.50/1

27.   a. 1.0 M $Al(NO_3)_3$

28.   0.41 mole

29.   basic from hydrolysis

30. $K_a \approx 1.5 \times 10^{-5}$

Chapter 12   ELECTROCHEMISTRY

Category I

1. a. $6 I^- + 8 H^+ + 2 NO_3^- \rightleftarrows 3 I_2 + 2 NO + 4 H_2O$
    b. $3 SO_2 + 2 H^+ + Cr_2O_7^{2-} \rightleftarrows 3 SO_4^{2-} + 2 Cr^{3+} + H_2O$
    c. $4 ClO_3^- \rightleftarrows 3 ClO_4^- + Cl^-$
    d. $2 F_2 + 2 H_2O \rightleftarrows O_2 + 4 H^+ + 4 F^-$
    e. $10 S_2O_3^{2-} + 16 H^+ + 2 MnO_4^- \rightleftarrows 5 S_4O_6^{2-} + 2 Mn^{2+} + 8 H_2O$
    f. $3 C_2H_5OH + 2 BrO_3^- \rightleftarrows 3 CH_3COOH + 2 Br^- + 3 H_2O$
    g. $2 CH_3CHO + 5 O_2 \rightleftarrows 4 CO_2 + 4 H_2O$

2. a. $Br^- + 2 H_2O + 2 CrO_4^{2-} \rightleftarrows BrO_3^- + Cr_2O_3 + 4 OH^-$
    b. $ClO^- + SO_3^{2-} \rightleftarrows Cl^- + SO_4^{2-}$
    c. $C_2O_4^{2-} + MnO_4^{2-} \rightleftarrows 2 CO_3^{2-} + MnO_2$
    d. $64 OH^- + 2 CrI_3 + 27 Cl_2 \rightleftarrows 2 CrO_4^{2-} + 6 IO_4^- + 32 H_2O + 54 Cl^-$
    e. $4 OH^- + 3 SnO + 2 CrO_4^{2-} + 11 H_2O \rightleftarrows 3 Sn(OH)_6^{2-} + 2 Cr(OH_4^-$
    f. $4 C_2H_5OH + BH_4^- \rightleftarrows 4 C_2H_6 + 2 H_2O + BO_2^-$
    g. $4 ClO_3^- \rightleftarrows 3 ClO_4^- + Cl^-$

3. $6 Fe^{2+} + Cr_2O_7^{2-} + 14 H^+ \rightleftarrows 6 Fe^{3+} + 2 Cr^{3+} + 7 H_2O$
Each mole of $Cr_2O_7^{2-}$ titrates 6 moles of Fe^{2+}.
moles $Cr_2O_7^{2-}$ = 0.50 M × 0.046 ℓ = 0.023, moles of Fe^{2+} = 6 × 0.023 = 0.138 mole

Category II

4.   (b)   anode: $Co \rightarrow Co^{2+} + 2 e^-$
            cathode: $Au^{3+} + 3 e^- \rightarrow Au$

(c) (Co) + 0.28 + (Au) + 1.50 = +1.78 V

(d) e^- flow from Co to Au; positive ion to Au, negative to Co.

5. Reaction: $PbO_2 + 4 H^+ + SO_4^{2-} + 2 e^- \rightarrow PbSO_4 + 2 H_2O$
From table of reduction potentials, E (standard) for anode reaction, ($Pb + SO_4^{2-} \rightarrow PbSO_4 + 2 e^-$) is +0.36 V, $\therefore$ potential for cathode reaction is +1.64 V.

6. $Li \rightarrow Li^+ + e^-$ and $F_2 + 2 e^- \rightarrow 2 F^-$ would produce 5.92 V. Both Li and F_2 react with H_2O and standard potentials are based on *aqueous* solutions. Perhaps there is another way?

7. (a) $Fe + Sn^{2+} \rightleftarrows Fe^{2+} + Sn$

(b) Fe anode, Sn cathode

(c) +0.44 (Fe) + -0.14 (Sn) = +0.30 V

8. (a) $H_2O \rightarrow O_2$ more readily than $NO_3^- \rightarrow ?$, Ag^+ more easily reduced than $H_2O \rightarrow H_2$.

(b) 0.75 amp $\times$ 35 min $\times$ 60 sec/min = 1575 coulombs

(c) $Xg\ Ag = 1575\ coul \times \dfrac{1\ mole\ e^-}{96,500\ coul} \times \dfrac{1\ mole\ Ag}{1\ mole\ e^-} \times \dfrac{108\ g\ Ag}{1\ mole\ Ag} = 1.76\ g$

9. $X\ sec = 8\ h \times \dfrac{60\ min}{1\ hr} \times \dfrac{60\ sec}{1\ min} = 28,800\ sec$

no. of coulombs = 20,000 amps $\times$ 28,800 sec = 5.76×10^8 coulombs

$X\ g\ Al = 5.76 \times 10^8\ coul \times \dfrac{1\ mole\ e^-}{96,500\ coul} \times \dfrac{1\ mole\ Al}{3\ moles\ e^-} \times \dfrac{27.0\ g\ Al}{1\ mole\ Al} = 5.37 \times 10^4\ g$

or 53.7 kg

10. $X\ coulombs = 327.0\ g\ Zn \times \dfrac{1\ mole\ Zn}{65.4\ g\ Zn} \times \dfrac{2\ moles\ e^-'s}{1\ mole\ Zn} \times \dfrac{96,500\ coulombs}{1\ mole\ e^-'s} =$

9.65×10^5 coulombs

$X\ sec = 9.65 \times 10^5\ coul/20.0\ amps = 4.83 \times 10^4\ sec$

$X\ h = 4.83 \times 10^4\ sec \times \dfrac{1\ min}{60\ sec} \times \dfrac{1\ hr}{60\ min} = 13.4\ h$

11. (a) coulombs = 5.0 amps $\times$ 16.1 min $\times$ 60 sec/min = 4830 coul

$X\ g\ Tc = 1\ mole\ e^-'s \times \dfrac{96,500\ coul}{1\ mole\ e^-'s} \times \dfrac{2.50\ g\ Tc}{4830\ coul} = 49.95\ g$

(b) Since at. wt. of Tc $\sim$99, the oxidation state must have been 2+.

Category III

12. (a) yes   (b) yes   (c) no   (d) no   (e) yes
(f) no   (g) yes

13. Standard potential appears to be

(1)	$2 Cu^+ \rightarrow 2 Cu^{2+} + 2 e^-$	−0.15 V
(2)	$Cu^{2+} + 2 e^- \rightarrow Cu$	+0.34 V
(3)	$2 Cu^+ \rightarrow Cu + Cu^{2+}$	+0.19 V

But, only one e^- is transferred in the final reaction. If we add free energies ($\Delta G = -23.06\ n\ E°$), $-23.06 \times 1 \times E_3^0 = (-23.06 \times 2 \times -0.15) + (-23.06 \times 2 \times +0.34)$ and $E_3^0 = +0.38$ V. This is the correct value, and $\Delta G = -23.06 \times 1 \times +0.38 = -8.76$ kcal; log K = 1 $\times 0.38/0.0592 = 6.42$; K = 2.63×10^6.

14. $E^0 = -0.77 + 0.80 = +0.03$ V
log K = 1 $\times +0.03/.0592 = 0.507$; K = 3.21

15. $E^0 = +0.28 - 0.25 = +0.03$ V for $Co + Ni^{2+} \rightleftarrows Co^{2+} + Ni$

log K = 2 × 0.03/.0592 = 1.0135; K = 10.3.

$$\frac{[Co^{2+}]}{[Ni^{2+}]} = \frac{(2.0 + X)}{(1.0 - X)} = 10.3 \; ; X = 0.73, \; [Ni^{2+}] = 0.27 \; M$$

16. E^0 = −0.44 + 0.41 = −0.03 V. ΔG = −23.06 × 2 × −0.03 = +1.38 kcal.
log K = 2 × (−0.03/.0592) = −1.0135
K = 9.7 × 10⁻²

17.   (a) E^0 = +0.76 − 0.14 = +0.62 V

(b) $E = E^0 - \frac{.0592}{n} \log \frac{(Zn^{2+})}{(Sn^{2+})}$ = +0.62 − (.03) × log (0.15/3.0)

E = +0.66 V

(c)   E = +0.62 − (.03) log (2.0/0.20) = +0.59 V

18. $E = E^0 - \frac{.0592}{n} \log \frac{[Pb^{2+}]}{[Cu^{2+}]}$; +0.55 = +0.47 − (.03) log $\frac{X}{1.0}$

X = [Pb²⁺] = 2.14 × 10⁻³. [Pb²⁺] [Cl⁻]² = Ksp = (2.14 × 10⁻³) (0.09)² = 1.7 × 10⁻⁵

General Problems

19.   7700 coulombs

20.   (b), (c), (d)

21.   (a)   8.56 × 10⁻⁶   (b)   8.55 × 10¹⁹

22.   (a)   S + 6, O − 2   (b)   H + 1, I + 7, O − 2   (c)   S + 2, O − 2
(d)   H + 1, P + 5, O − 2   (e)   H + 1, Br + 1, O − 2   (f)   N + 3, O − 2

23.   K = 4 × 10³ ; E^0 = +0.071 V; ΔG⁰ = −4.9 kcal

24.   −0.34 V

Chapter 13   NUCLEAR REACTIONS

Category I

1.   (a) $^{36}_{17}Cl \rightarrow \, ^{36}_{18}Ar + \, ^{0}_{-1}e$
(b) $^{13}_{7}N \rightarrow \, ^{13}_{6}C + \, ^{0}_{1}e$
(c) $^{230}_{90}Th \rightarrow \, ^{226}_{88}Ra + \, ^{4}_{2}He$

2. $^{59}_{27}Co + \, ^{2}_{1}H \rightarrow \, ^{60}_{27}Co + \, ^{1}_{1}H$
$^{60}_{27}Co \rightarrow \, ^{60}_{28}Ni + \, ^{0}_{-1}e$

3.   (a)   beta   (b)   positron   (c)   positron   (d)   beta

4. $^{37}_{17}Cl + \, ^{2}_{1}H \rightarrow \, ^{35}_{16}S + \, ^{4}_{2}He$
$^{208}_{82}Pb + \, ^{4}_{2}He \rightarrow \, ^{210}_{84}Po + 2 \, ^{1}_{0}n$
$^{44}_{20}Ca + \, ^{2}_{1}H \rightarrow \, ^{45}_{20}Ca + \, ^{1}_{1}H$
$^{209}_{83}Bi + \, ^{2}_{1}H \rightarrow \, ^{210}_{84}Po + \, ^{1}_{0}n$

5. $^{220}_{86}Rn \rightarrow \, ^{216}_{84}Po + \, ^{4}_{2}He; \; ^{216}_{84}Po \rightarrow \, ^{212}_{82}Pb + \, ^{4}_{2}He;$
$^{212}_{82}Pb \rightarrow \, ^{212}_{83}Bi + \, ^{0}_{-1}e; \; ^{212}_{83}Bi \rightarrow \, ^{208}_{81}Tl + \, ^{4}_{2}He;$
$^{208}_{81}Tl \rightarrow \, ^{208}_{82}Pb + \, ^{0}_{-1}e$ (one possibility)

Category II

6. $\log \frac{15,000}{11,905} = \frac{k(5.0 \; hrs)}{2.30}$; k = 4.62 × 10⁻² hrs⁻¹ ; $t_{1/2}$ = 0.693/k = 15.0 hrs

7. $k = 1.20 \times 10^{-4}$ years^{-1}

$$\log \frac{100}{59.55} = \frac{(1.20 \times 10^{-4} \text{ yrs}^{-1}) \times t}{2.30} \; ; t = 4313 \text{ years}$$

8. $k = 0.258$ days^{-1}

$$\log \frac{100}{5} = \frac{(0.258 \text{ days}^{-1}) \times t}{2.30} \; ; t = 11.6 \text{ days}$$

9. For ^{40}K, $k = 5.8 \times 10^{-10}$ yrs^{-1} (see Example 2). Rate equation shows 92.6% of ^{40}K decayed in 4.5×10^9 yrs. If half that amount has decayed on Mars, then 53.7% remains, and

$$\log \frac{100}{53.7} = \frac{(5.8 \times 10^{-10} \text{ yrs}^{-1}) \times t}{2.30} \; ; t = 1.1 \times 10^9 \text{ years}$$

But, ^{40}Ar may readily escape from rocks in the low atmospheric pressure of Mars.

10. $t_{\frac{1}{2}} = 138$ days $= 1.19 \times 10^7$ sec; $k = 5.8 \times 10^{-8}$ sec^{-1}

Number of atoms needed to give decay of 3.7×10^{10} atoms/sec $= 3.7 \times 10^{10}$ atoms/sec $\div 5.8 \times 10^{-8}$ sec$^{-1} = 6.4 \times 10^{17}$.

$$\text{Xg} \; ^{210}\text{Po} = 6.4 \times 10^{17} \text{ atoms} \times \frac{210 \text{ g } ^{210}\text{Po}}{6.02 \times 10^{23} \text{ atoms}} = 2.2 \times 10^{-4} \text{ g, or } 0.22 \text{ mg}$$

11. $k = 0.693/12.46$ yrs $= 5.56 \times 10^{-2}$ yrs^{-1}

(a) $\log \dfrac{2.00 \text{ g}}{X} = \dfrac{(5.56 \times 10^{-2} \text{ yrs}^{-1}) \times (5 \text{ yrs})}{2.30} \; ; X = 1.51 \text{ g}$

(b) For $t = 15$ yrs, $X = 0.868$ g

(c) For $t = 25$ yrs, $X = 0.498$ g

Category III

12. (a) True mass $= 4.00150$ g per mole

Calculated mass $= (2 \times 1.00728) + (2 \times 1.00867) = 4.03190$ g

$\Delta E = (0.03040 \text{ g}) \times (2.15 \times 10^{10} \text{ kcal}) = 6.54 \times 10^8$ kcal

(b) True mass $= 39.95162$ g

Calculated mass $= (20 \times 1.00728) + (20 \times 1.00867) = 40.31900$ g

$\Delta E = (0.3674 \text{ g}) \times (2.15 \times 10^{10} \text{ kcal/g}) = 7.90 \times 10^9$ kcal

(c) True mass $= 205.9295$

Calculated mass $= (82 \times 1.00728) + (124 \times 1.00867) = 207.6720$

$\Delta E = (1.7425 \text{ g}) \times (2.15 \times 10^{10} \text{ kcal/g}) = 3.75 \times 10^{10}$ kcal

13. (a) $\Delta m = 3.01550 - (3.01493 + 0.000549) = 2.1 \times 10^{-5}$ g/mole

$(2.1 \times 10^{-5}) \times (2.15 \times 10^{10} \text{ kcal/g}) = 4.52 \times 10^5$ kcal/mole

(b) $\Delta m = (11.00656 + 2.01355) - (11.00814 + 2 \times 1.00867) = -5.37 \times 10^{-3}$

(endothermic reaction)

$(5.37 \times 10^{-3}) \times (2.15 \times 10^{10} \text{ kcal/g}) = 1.15 \times 10^8$ kcal/mole

(c) $\Delta m = (238.0003) - (233.9934 + 4.00150) = 5.4 \times 10^{-3}$

$(5.4 \times 10^{-3}) \times (2.15 \times 10^{10}) = 1.16 \times 10^8$ kcal/mole

14. (a) Mass loss – per mole for $(4 \, ^1_1\text{H} \rightarrow \, ^4_2\text{He} + 2 \, ^0_1\text{e}) = (4.00150 + 2 \times .000549) - (4 \times 1.00728) = -0.02652$ grams

$$X \text{ kcal} = 0.02652 \text{ g} \times \frac{2.15 \times 10^{10} \text{ kcal}}{1 \text{ g}} = 5.70 \times 10^8 \text{ kcal}$$

Mass loss per day:

$$X \text{ g} = 1 \text{ day} \times \frac{24 \text{ hrs}}{1 \text{ day}} \times \frac{3600 \text{ sec}}{1 \text{ hr}} \times \frac{9.56 \times 10^{22} \text{ kcal}}{1 \text{ sec}} \times \frac{0.02652 \text{ g}}{5.70 \times 10^8 \text{ kcal}} =$$

3.84×10^{17} g

(b) $1/10^6 \times 2.0 \times 10^{33}$ kg = 2.0×10^{27} kg = 2.0×10^{30} g

$\qquad$ X yrs = 2.0×10^{30} g $\times \dfrac{1 \text{ day}}{3.84 \times 10^{17} \text{ g}} \times \dfrac{1 \text{ yr}}{365 \text{ days}}$ = 1.4×10^{10} yrs

15.   Total mass = $2 \times 1.00728 + 2 \times 0.000549$ = 2.01566 g

$\qquad$ X kcal = 2.01566 g $\times \dfrac{2.15 \times 10^{10} \text{ kcal}}{1 \text{ g}}$ = 4.33×10^{10} kcal

General Problems

16.   (a) $^2_1$H $\qquad$ (b) $^4_2$He $\qquad$ (c) $^1_1$H

17.   1.72 hrs

18. 1.0×10^{-11} g

19. 3.85×10^8 kcal/mole

20. 9.7×10^{18} emissions per second

APPENDIX I
MATH PRACTICE

Often, beginning chemistry students are known to proclaim, "It's not chemistry that bothers me in this course, it's the math!" Relax. The math in this book and the great majority of beginning chemistry courses consists only of manipulating exponential numbers, logarithms and some simple algebra.

Exponential numbers

Writing numbers to powers of 10 is just a convenient way of handling very large or very small numbers. How much simpler it is to write 6.02×10^{23} than it is to write out 602,000,000,000,000,000,000,000. Or to say a solution is 1.52×10^{-6} M in H^+, rather than 0.00000152 M in H^+.

Students sometimes have trouble multiplying and dividing exponential numbers. Do the exponents add or subtract? Which way should we move the decimal point? One useful trick to avoid confusion is to translate the *type* of problem into a *similar* but simple problem. Let's illustrate. Examine the simple problem on the right, then solve the problem on the left. Notice how you handle the exponents.

	PROBLEM	*SIMPLIFIED PROBLEM*
(1)	$(6.0 \times 10^{23}) \times (2.0 \times 10^2) = ?$	$(6.0 \times 10^2) \times (2.0 \times 10^1) = 600 \times 20 = 12,000 = 12.0 \times 10^3 = 1.20 \times 10^4.$
(2)	$(3.2 \times 10^{-8}) \times (4.0 \times 10^3) = ?$	$(3.2 \times 10^{-2}) \times (4.0 \times 10^1) = .032 \times 40 = 1.28 = 12.8 \times 10^{-1} = 1.28 \times 10^0.$
(3)	$(6.6 \times 10^{-7}) \times (4.1 \times 10^{-2}) = ?$	$(6.6 \times 10^{-2}) \times (4.1 \times 10^{-1}) = .066 \times .41 = .027 = 27.0 \times 10^{-3} = 2.70 \times 10^{-2}.$
(4)	$\dfrac{5.5 \times 10^{16}}{2.1 \times 10^4} = ?$	$\dfrac{5.5 \times 10^2}{2.1 \times 10^1} = \dfrac{550}{21} = 26.2 = 2.62 \times 10^1$
(5)	$\dfrac{1.0 \times 10^{-14}}{4.3 \times 10^{-5}} = ?$	$\dfrac{1.0 \times 10^{-2}}{4.3 \times 10^{-1}} = \dfrac{.01}{.43} = .023 = 0.23 \times 10^{-1} = 2.3 \times 10^{-2}$
(6)	$\dfrac{7.2 \times 10^{18}}{2.3 \times 10^{-5}} = ?$	$\dfrac{7.2 \times 10^2}{2.3 \times 10^{-1}} = \dfrac{720}{.23} = 3130 = 3.13 \times 10^3$
(7)	$\dfrac{1.92 \times 10^{-6}}{8.3 \times 10^4} = ?$	$\dfrac{1.92 \times 10^{-2}}{8.3 \times 10^1} = \dfrac{.0192}{83} = .00023 = 0.23 \times 10^{-3} = 2.3 \times 10^{-4}$
(8)	$(2.2 \times 10^{-5})^2 = ?$	$(2.2 \times 10^{-1})^2 = .22 \times .22 = .0484 = 4.84 \times 10^{-2}$

175

When multiplying, the exponents add. When dividing, the exponents subtract. When deciding which way to "move" the decimal point, setting up a simplified version of the problem can be very helpful. If you were able to solve the problems on the left you are ready for any chemistry problem involving exponential numbers. The answers are (1) 1.20×10^{26} (2) $12.8 \times 10^{-5} = 1.28 \times 10^{-4}$ (3) $27.1 \times 10^{-9} = 2.71 \times 10^{-8}$ (4) 2.62×10^{12} (5) $0.23 \times 10^{-9} = 2.3 \times 10^{-10}$ (6) 3.13×10^{23} (7) $0.23 \times 10^{-10} = 2.3 \times 10^{-11}$ (8) 4.84×10^{-10}

For practice, do the following. *Multiply*: (1) 4.02×10^{-6} by 2.61×10^{-3} (2) 7.2×10^{16} by 4.2×10^{-5} (3) 9.9×10^{8} by 2.3×10^{4} (4) 4000 by 6.0×10^{-3} (5) 62.3×10^{-5} by 2.0×10^{-2}. *Divide*: (6) 6.02×10^{23} by 9.2×10^{-5} (7) 4000 by 6.0×10^{-3} (8) 1.0×10^{-14} by 7.6×10^{-4} (9) 108.0×10^{2} by 0.42×10^{-4} (10) 6.6×10^{8} by 7.0×10^{3} (11) 8.8×10^{-3} by 2.2×10^{5} (12) 2.7×10^{-6} by .082

Answers: (1) 1.05×10^{-8} (2) 3.0×10^{12} (3) 2.3×10^{13} (4) 24 (5) 1.25×10^{-5} (6) 6.5×10^{27} (7) 6.7×10^{5} (8) 1.3×10^{-11} (9) 2.6×10^{8} (10) 9.4×10^{4} (11) 4.0×10^{-8} (12) 3.3×10^{-5}

Logarithms

Logarithms are just exponents. A common logarithm to base 10 is the power to which 10 must be raised to give a desired number. For example:

(1) $\log 1000 = \log 10^3 = 3$
(2) $\log 0.01 = \log 10^{-2} = -2$
(3) $\log 2000 = \log 2 \times 10^3 = \log 2 + \log 10^3 = .301 + 3 = 3.301$
(4) $\log 0.02 = \log 2 \times 10^{-2} = \log 2 + \log 10^{-2} = .301 + (-2) = -1.699$

Examples (1) and (2) are straightforward. Examples (3) and (4) are not quite so obvious. First, note that for any number, $A \times B$, the log of $A \times B = \log A + \log B$. This is true because logarithms are exponents and when multiplying, exponents add. Thus $\log 2 \times 10^3$ is equal to $\log 2 + \log 10^3$. The log of 10^3 is just 3, but how do we find the log of 2? To what power must 10 be raised to give 2? This is not an easy operation to carry out "long hand." Fortunately tables of logarithms are readily available and we can look up the log of 2 (see p. 179). We find the log of 2 (or 2.0) equals .301. This means that $10^{.301} = 2$. Then, in example (3), we simply add .301 + 3 to give 3.301. This is the log of 2×10^3. The log of 2×10^{-2}, in example (4), is obtained in a similar fashion. Check and make sure you understand the following:

$$\log 3.0 \times 10^4 = 4.477$$
$$\log 2.2 \times 10^6 = 6.342$$
$$\log 5.47 \times 10^3 = 3.738$$
$$\log 6.0 \times 10^{-5} = -4.222$$
$$\log 7.62 \times 10^{-9} = -8.118$$

When dividing, exponents are subtracted. Thus, the log of $\dfrac{A}{B}$ is the same as $\log A - \log B$. If we have a problem such as $\log \dfrac{4.0 \times 10^8}{2.0 \times 10^2} = ?$, we can proceed by carrying out the division first to give $\log 2.0 \times 10^6 = ?$ (ans. 6.301), or we could find $(\log 4.0 \times 10^8) - (\log 2.0 \times 10^2)$. For the latter we find: $8.602 - 2.301 = 6.301$.

If we have a logarithm and need the number that corresponds to it (the antilog), we simply reverse the process. If we know the log of a number is .895, we look for that log in the table and find the number having that logarithm (7.85). Check the following:

log	number
1.949	8.9×10^1
-6.462	3.45×10^{-7}
-2.308	4.92×10^{-3}
0.0682	1.17
5.734	5.42×10^5

In this book we use logarithms in three chapters: Nuclear Reactions ($\log \frac{X_0}{X} = \frac{kt}{2.30}$), Rates of Reaction ($\log \frac{X_0}{X} = \frac{kt}{2.30}$ and $\log \frac{k_2}{k_1} = \frac{Ea}{4.58} \left(\frac{T_2 - T_1}{T_2 T_1} \right)$), and Acid-Base Equilibria (pH = $-\log$ [H^+]). In the discussion of pH, a shortcut method is introduced. That is, when [H^+] = A $\times$ 10^{-B}, and the pH is B $-$ log A. Let's see why that shortcut works. Suppose [H^+] = 5.35×10^{-4}. Then pH = $-\log (5.35 \times 10^{-4})$ = $-(\log 5.35 + \log 10^{-4})$ = $-(.728 + -4)$ = $(+4 - .728)$ = 3.272. Hence, the pH is just 4 $-$ log 5.35.

APPENDIX II
REFERENCE TABLES

I. Constants

Absolute zero of temperature	$= 0°K = -273.16°C$
Avogadro's number	$= 6.022 \times 10^{23}$
Ideal Gas constant	$= 0.08206$ liter-atm/mole-$°K$
	$= 1.987$ cal/mole-$°K$
Velocity of light	$= 2.998 \times 10^{10}$ cm/sec
Electron charge	$= 1.602 \times 10^{-19}$ coulomb
Faraday constant	$= 96,487$ joule/volt $= 23.06$ kcal/volt
ln X/log X	$= 2.303$

II. Conversion Factors

Length
- 1 meter = 100 cm = 1000 mm = 39.37 inches
- 1 inch = 2.54 cm
- 1 foot = 30.48 cm
- 1 angstrom = 10^{-8} cm

Mass
- 1 kilogram = 1000 grams = 2.20 pounds
- 1 pound = 453.6 grams
- 1 gram = 0.03527 ounce

Volume
- 1 liter = 1.057 quarts
- 1 ml = 1 cc (cm^3) = 10^{-3} liter

Energy
- 1 kcal = 4184 joules
- 1 joule = 10^7 ergs = 6.33×10^{18} electron volts
- 1 cal = 4.129×10^{-2} liter-atm

Temperature
- $°F = 1.8°C + 32$
- $°K = °C + 273$

Pressure
- 1 atm = 760 mm Hg = 760 Torr = 1.01325×10^5 pascals (Pa)
- 1 bar = 10^5 Pa
- 1 lb/in^2 = 6895 Pa

III. Other Tables in Text

IV. Table of Logarithms

	0	1	2	3	4	5	6	7	8	9
1.0	.0000	.0043	.0086	.0128	.0170	.0212	.0253	.0294	.0334	.0374
1.1	.0414	.0453	.0492	.0531	.0569	.0607	.0645	.0682	.0719	.0755
1.2	.0792	.0828	.0864	.0899	.0934	.0969	.1004	.1038	.1072	.1106
1.3	.1139	.1173	.1206	.1239	.1271	.1303	.1335	.1367	.1399	.1430
1.4	.1461	.1492	.1523	.1553	.1584	.1614	.1644	.1673	.1703	.1732
1.5	.1761	.1790	.1818	.1847	.1875	.1903	.1931	.1959	.1987	.2014
1.6	.2041	.2068	.2095	.2122	.2148	.2175	.2201	.2227	.2253	.2279
1.7	.2304	.2330	.2355	.2380	.2405	.2430	.2455	.2480	.2504	.2529
1.8	.2553	.2577	.2601	.2625	.2648	.2672	.2695	.2718	.2742	.2765
1.9	.2788	.2810	.2833	.2856	.2878	.2900	.2923	.2945	.2967	.2989
2.0	.3010	.3032	.3054	.3075	.3096	.3118	.3139	.3160	.3181	.3201
2.1	.3222	.3243	.3263	.3284	.3304	.3324	.3345	.3365	.3385	.3404
2.2	.3424	.3444	.3464	.3483	.3502	.3522	.3541	.3560	.3579	.3598
2.3	.3617	.3636	.3655	.3674	.3692	.3711	.3729	.3747	.3766	.3784
2.4	.3802	.3820	.3838	.3856	.3874	.3892	.3909	.3927	.3945	.3962
2.5	.3979	.3997	.4014	.4031	.4048	.4065	.4082	.4099	.4116	.4133
2.6	.4150	.4166	.4183	.4200	.4216	.4232	.4249	.4265	.4281	.4298
2.7	.4314	.4330	.4346	.4362	.4378	.4393	.4409	.4425	.4440	.4456
2.8	.4472	.4487	.4502	.4518	.4533	.4548	.4564	.4579	.4594	.4609
2.9	.4624	.4639	.4654	.4669	.4683	.4698	.4713	.4728	.4742	.4757
3.0	.4771	.4786	.4800	.4818	.4829	.4843	.4857	.4871	.4886	.4900
3.1	.4914	.4928	.4942	.4955	.4969	.4983	.4997	.5001	.5024	.5038
3.2	.5051	.5065	.5079	.5092	.5105	.5119	.5132	.5145	.5159	.5172
3.3	.5185	.5198	.5211	.5224	.5237	.5250	.5263	.5276	.5289	.5302
3.4	.5315	.5328	.5340	.5353	.5366	.5378	.5391	.5403	.5416	.5428
3.5	.5441	.5453	.5465	.5478	.5490	.5502	.5514	.5527	.5539	.5551
3.6	.5563	.5575	.5587	.5599	.5611	.5623	.5635	.5647	.5658	.5670
3.7	.5682	.5694	.5705	.5717	.5729	.5740	.5752	.5763	.5775	.5786
3.8	.5798	.5809	.5821	.5832	.5843	.5855	.5866	.5877	.5888	.5899
3.9	.5911	.5922	.5933	.5944	.5955	.5966	.5977	.5988	.5999	.6010
4.0	.6021	.6031	.6042	.6053	.6064	.6075	.6085	.6096	.6107	.6117
4.1	.6128	.6138	.6149	.6160	.6170	.6180	.6191	.6201	.6212	.6222
4.2	.6232	.6243	.6253	.6263	.6274	.6284	.6294	.6304	.6314	.6325
4.3	.6335	.6345	.6355	.6365	.6375	.6385	.6395	.6405	.6414	.6425
4.4	.6435	.6444	.6454	.6464	.6474	.6484	.6493	.6503	.6513	.6522
4.5	.6532	.6542	.6551	.6561	.6571	.6580	.6590	.6599	.6609	.6618
4.6	.6628	.6637	.6646	.4456	.6665	.6675	.6684	.6693	.6702	.6712
4.7	.6721	.6730	.6739	.6749	.6758	.6767	.6776	.6785	.6794	.6803
4.8	.6812	.6821	.6830	.6839	.6848	.6857	.6866	.6875	.6884	.6893
4.9	.6902	.6911	.6920	.6928	.6937	.6946	.6955	.6964	.6972	.6981
5.0	.6990	.6998	.7007	.7016	.7024	.7033	.7042	.7050	.7059	.7067
5.1	.7076	.7084	.7093	.7101	.7110	.7118	.7126	.7135	.7143	.7152
5.2	.7160	.7168	.7177	.7185	.7193	.7202	.7210	.7218	.7226	.7235
5.3	.7243	.7251	.7259	.7267	.7275	.7284	.7292	.7300	.7308	.7316
5.4	.7324	.7332	.7340	.7348	.7356	.7364	.7372	.7380	.7388	.7396
5.5	.7404	.7412	.7419	.7427	.7435	.7443	.7451	.7459	.7466	.7474
5.6	.7482	.7490	.7497	.7505	.7513	.7520	.7528	.7536	.7543	.7551
5.7	.7559	.7566	.7574	.7582	.7589	.7597	.7604	.7612	.7619	.7627
5.8	.7634	.7642	.7649	.7657	.7664	.7672	.7679	.7686	.7694	.7701
5.9	.7709	.7716	.7723	.7731	.7738	.7745	.7752	.7760	.7767	.7774

Table of Logarithms (Continued)

	0	1	2	3	4	5	6	7	8	9
6.0	.7782	.7789	.7796	.7803	.7810	.7818	.7825	.7832	.7839	.7846
6.1	.7853	.7860	.7868	.7875	.7882	.7889	.7896	.7903	.7910	.7917
6.2	.7924	.7931	.7938	.7945	.7952	.7959	.7966	.7973	.7980	.7987
6.3	.7993	.8000	.8007	.8014	.8021	.8028	.8035	.8041	.8048	.8055
6.4	.8062	.8069	.8075	.8082	.8089	.8096	.8102	.8109	.8116	.8122
6.5	.8129	.8136	.8142	.8149	.8156	.8162	.8169	.8176	.8182	.8189
6.6	.8195	.8202	.8209	.8215	.8222	.8228	.8235	.8241	.8248	.8254
6.7	.8261	.8267	.8274	.8280	.8287	.8293	.8299	.8306	.8312	.8319
6.8	.8325	.8331	.8338	.8344	.8351	.8357	.8363	.8370	.8376	.8382
6.9	.8388	.8395	.8401	.8407	.8414	.8420	.8426	.8432	.8439	.8445
7.0	.8451	.8457	.8463	.8470	.8476	.8482	.8488	.8495	.8500	.8506
7.1	.8513	.8519	.8525	.8531	.8537	.8543	.8549	.8555	.8561	.8567
7.2	.8573	.8579	.8585	.8591	.8597	.8603	.8609	.8615	.8621	.8627
7.3	.8633	.8639	.8645	.8651	.8657	.8663	.8669	.8675	.8681	.8686
7.4	.8692	.8698	.8704	.8710	.8716	.8722	.8727	.8733	.8739	.8745
7.5	.8751	.8756	.8762	.8768	.8774	.8779	.8785	.8791	.8797	.8802
7.6	.8808	.8814	.8820	.8825	.8831	.8837	.8842	.8848	.8854	.8859
7.7	.8865	.8871	.8876	.8882	.8887	.8893	.8899	.8904	.8910	.8915
7.8	.8921	.8927	.8932	.8938	.8943	.8949	.8954	.8960	.8965	.8971
7.9	.8976	.8982	.8987	.8993	.8998	.9004	.9009	.9015	.9020	.9026
8.0	.9031	.9036	.9042	.9047	.9053	.9058	.9063	.9069	.9074	.9079
8.1	.9085	.9090	.9096	.9101	.9106	.9112	.9117	.9122	.9128	.9133
8.2	.9138	.9143	.9149	.9154	.9159	.9165	.9170	.9175	.9180	.9186
8.3	.9191	.9196	.9201	.9206	.9212	.9217	.9222	.9227	.9232	.9238
8.4	.9243	.9248	.9253	.9258	.9263	.9269	.9274	.9279	.9284	.9289
8.5	.9294	.9299	.9304	.9309	.9315	.9320	.9325	.9330	.9335	.9340
8.6	.9345	.9350	.9355	.9360	.9365	.9370	.9375	.9380	.9385	.9390
8.7	.9395	.9400	.9405	.9410	.9415	.9420	.9425	.9430	.9435	.9440
8.8	.9445	.9450	.9455	.9460	.9465	.9469	.9474	.9479	.9484	.9489
8.9	.9494	.9499	.9504	.9509	.9513	.9518	.9523	.9528	.9533	.9538
9.0	.9542	.9547	.9552	.9557	.9562	.9566	.9571	.9576	.9581	.9586
9.1	.9590	.9595	.9600	.9605	.9609	.9614	.9619	.9624	.9628	.9633
9.2	.9638	.9643	.9647	.9652	.9657	.9661	.9666	.9671	.9675	.9680
9.3	.9685	.9689	.9694	.9699	.9703	.9708	.9713	.9717	.9722	.9727
9.4	.9731	.9736	.9741	.9745	.9750	.9754	.9759	.9763	.9768	.9773
9.5	.9777	.9782	.9786	.9791	.9795	.9800	.9805	.9809	.9814	.9818
9.6	.9823	.9827	.9832	.9836	.9841	.9845	.9850	.9854	.9859	.9863
9.7	.9868	.9872	.9877	.9881	.9886	.9890	.9894	.9899	.9903	.9908
9.8	.9912	.9917	.9921	.9926	.9930	.9934	.9939	.9943	.9948	.9952
9.9	.9956	.9961	.9965	.9969	.9974	.9978	.9983	.9987	.9991	.9996

INDEX